STUDENT'S SOLUTIONS MANUAL

VOLUME 1: CHAPTERS 1–16

THIRD EDITION

college
a strategic approach
physics

knight • jones • field

LARRY K. SMITH
SNOW COLLEGE

PEARSON

Boston Columbus Indianapolis New York San Francisco Hoboken
Amsterdam Cape Town Dubai London Madrid Milan Munich Paris Montréal Toronto
Delhi Mexico City São Paulo Sydney Hong Kong Seoul Singapore Taipei Tokyo

Executive Editor: Becky Ruden

Project Manager: Martha Steele

Managing Development Editor: Cathy Murphy

Senior Development Editor: Alice Houston

Editorial Assistant: Sarah Kaubisch

Compositor: PreMediaGlobal, Inc.

Cover Designer: Tandem Creative, Inc. / Seventeenth Street Studios

Manufacturing Buyer: Jeff Sargent

Vice-President of Marketing: Christy Lesko

Marketing Manager: Will Moore

Senior Market Development Manager: Michelle Cadden

Cover Photo Credit: Franklin Kappa

Credits and acknowledgments borrowed from other sources and reproduced, with permission, in this textbook appear on the appropriate page within the text.

Many of the designations used by manufacturers and sellers to distinguish their products are claimed as trademarks. Where those designations appear in this book, and the publisher was aware of a trademark claim, the designations have been printed in initial caps or all caps.

MasteringPhysics® is a trademark, in the U.S. and/or other countries, of Pearson Education, Inc. or its affiliates.

ISBN 10: 0-321-90884-8

ISBN 13: 978-0-321-90884-1

PEARSON   www.pearsonhighered.com

1 2 3 4 5 6 7 8 9 10—DOH—15 14

Table of Contents

Preface

This *Student's Solutions Manual* is intended to provide you with examples of good problem-solving techniques and strategies. To achieve that, the solutions presented here attempt to:

- Follow, in detail, the problem-solving strategies presented in the text.
- Articulate the reasoning that must be done before computation.
- Illustrate how to use drawings effectively.
- Demonstrate how to utilize graphs, ratios, units, and the many other "tactics" that must be successfully mastered and marshaled if a problem-solving strategy is to be effective.
- Show examples of assessing the reasonableness of a solution.
- Comment on the significance of a solution or on its relationship to other problems.

We recommend you try to solve each problem on your own before you read the solution. Simply reading solutions, without first struggling with the issues, has limited educational value.

As you work through each solution, make sure you understand how and why each step is taken. See if you can understand which aspects of the problem made this solution strategy appropriate. You will be successful on exams not by memorizing solutions to particular problems but by coming to recognize which kinds of problem-solving strategies go with which types of problems.

We have made every effort to be accurate and correct in these solutions. However, if you do find errors or ambiguities, we would be very grateful to hear from you.

REPRESENTING MOTION

Q1.1. Reason: (a) The basic idea of the particle model is that we will treat an object *as if* all its mass is concentrated into a single point. The size and shape of the object will not be considered. This is a reasonable approximation of reality if: (i) the distance traveled by the object is large in comparison to the size of the object, and (ii) rotations and internal motions are not significant features of the object's motion. The particle model is important in that it allows us to *simplify* a problem. Complete reality—which would have to include the motion of every single atom in the object—is too complicated to analyze. By treating an object as a particle, we can focus on the most important aspects of its motion while neglecting minor and unobservable details.
(b) The particle model is valid for understanding the motion of a satellite or a car traveling a large distance.
(c) The particle model is not valid for understanding how a car engine operates, how a person walks, how a bird flies, or how water flows through a pipe.
Assess: Models are representations of reality—not reality itself. As such they almost all make some simplifying assumptions. The test of a good model is the results it produces. The particle model allows us to model the motion of many objects simply and see common features of the movement of different objects. When used appropriately it is very useful. When used outside the range of its validity, it isn't very helpful.

Q1.5. Reason: Position refers to the location of an object at a given time relative to a coordinate system. Displacement, on the other hand, is the difference between the object's final position at time t_f and the initial position at time t_i. Displacement is a vector, whose direction is from the initial position toward the final position. An airplane at rest relative to a runway lamp, serving as the origin of our coordinate system, will have a position, called the initial position. The location of the airplane as it takes off may be labeled as the final position. The difference between the two positions, final minus initial, is displacement.
Assess: Some physics texts are not as explicit or clear about this terminology, but it pays off to have clear definitions for terms and to use them consistently.

Q1.9. Reason: If the position of the bicycle is negative it is to your left. The bicycle's velocity is positive, or to the right, so the bicycle is getting closer to you.
Assess: If the initial position had been positive and the velocity positive, the bicycle would be getting farther away from you.

Q1.11. Reason:

Start

Assess: The dots get farther apart and the velocity arrows get longer as she speeds up.

Q1.15. Reason:

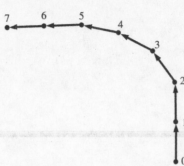

Assess: The car (particle) moves at a constant speed v so the distance between the dots is constant. While turning v remains constant, but the direction of $\vec{v}$ changes.

Q1.17. Reason: Because density is defined to be the ratio of two scalars (mass and volume), it too must be a scalar.
Assess: This is always true. The ratio of scalars is a scalar. On the other hand, a vector divided by a scalar (as in the definition of velocity) is a vector.

Q1.19. Reason: Since the rock is above the origin the position is positive; since it is still moving upward the velocity is also positive. Hence, the correct answer is A.
Assess: After it gets to the top and starts back down, the position will still be positive, but the velocity will be negative.

Q1.23. Reason: The speed is the distance divided by the time.

$$\text{speed} = \frac{\text{distance}}{\text{time}} = \frac{0.30 \text{ km}}{5.0 \text{ min}} \left(\frac{1000 \text{ m}}{1 \text{ km}} \right) \left(\frac{1 \text{ min}}{60 \text{ s}} \right) = 1.0 \text{ m/s}$$

So the correct choice is B.
Assess: 1 m/s does seem like a reasonable speed for a seal in water.

Q1.25. Reason: We are given the person's speed, so we can calculate the distance traveled in a given time using Equation 1.1. However, we are given the speed in meters/second, the distance in miles, and the time in minutes. We must convert units to be consistent. We will convert to SI units since they are standard in science. 1 mile $= 1.609 \times 1000$ m $= 1609$ m. Now we can apply Equation 1.1:

$$\text{time} = \frac{\text{distance}}{\text{speed}} = \frac{1609 \text{ m}}{2.00 \text{ m/s}} = 8.05 \times 10^2 \text{ s}$$

Note how the unit of meters cancels in the above equation.
Now we need to convert seconds to minutes.

$$(8.05 \times 10^2 \text{ s}) \left(\frac{1 \text{ min}}{60 \text{ s}} \right) = 13.4 \text{ minutes}$$

The correct choice is B.
Assess: It's important to use consistent units in a problem. After canceling like units, the remaining units should be the units you expect in the answer.

Q1.29. Reason: We are given an equation for density and are asked to calculate the density of the earth given its mass and volume. However, the units must be converted before the calculation is done since we're given volume in km^3 and the answer must be given in terms of m^3.

$$V = (1.08 \times 10^{12} \ \text{km}^3) \left(\frac{10^3 \ \text{m}}{1 \ \text{km}} \right) \left(\frac{10^3 \ \text{m}}{1 \ \text{km}} \right) \left(\frac{10^3 \ \text{m}}{1 \ \text{km}} \right)$$

$$= (1.08 \times 10^{12} \ \text{km}^3) \left(\frac{10^3 \ \text{m}}{1 \ \text{km}} \right)^3 = (1.08 \times 10^{12} \ \text{km}^3) \left(\frac{10^9 \ \text{m}^3}{1 \ \text{km}^3} \right)$$

$$= 1.08 \times 10^{21} \ \text{m}^3$$

Note carefully that we needed *three* conversion factors for the conversion from km to m here since we are dealing with cubic kilometers. Three factors are needed to cancel the factor of $\text{km}^3 = \text{km} \cdot \text{km} \cdot \text{km}$.

So the density is

$$\rho = \frac{M}{V} = \frac{5.94 \times 10^{24} \ \text{kg}}{1.08 \times 10^{21} \ \text{m}^3} = 5.50 \times 10^3 \ \text{kg/m}^3$$

The correct choice is A.

Assess: For cubic and square units (or units to any power) you must include the correct number of conversion factors to convert every factor in the original quantity. Since the density of water is $1.0 \times 10^3 \ \text{kg/m}^3$, it seems reasonable that the earth would be 5.5 times as dense.

Problems

P1.1. Prepare: Frames of the video are taken at equal intervals of time. As a result, we have a record of the position of the car at successive time equal intervals – this information allows us to construct a motion diagram.
Solve:

Skid
begins

Stops

Assess: Once the brakes are applied, the car slows down and travels a smaller distance during each successive time interval until it stops. This is what the car in the figure is doing.

P1.5. Prepare: Displacement is the difference between a final position x_f and an initial position x_i This can be written as $\Delta x = x_f - x_i$, and we are given that $x_i = 23 \ \text{m}$ and that $\Delta x = -45 \text{m}$.
Solve: $\Delta x = x_f - x_i$
Since we want to know the final position we solve this for x_f.

$$x_f = x_i + \Delta x$$
$$= 23 \ \text{m} + (-45 \ \text{m})$$
$$= -22 \ \text{m}$$

Assess: A negative displacement means a movement to the left, and Keira has moved left from $x = 23 \ \text{m}$ to $x = -22 \ \text{m}$.

P1.7. Prepare: We have been given three different displacements. The problem is straightforward since all the displacements are along a straight east-west line. All we have to do is add the displacements and see where we end up.
Solve: The first displacement is $\Delta \vec{x}_1 = 500 \ \text{m}$ east, the second is $\Delta \vec{x}_2 = 400 \ \text{m}$ west and the third displacement is $\Delta \vec{x}_3 = 700 \ \text{m}$ east. These three displacements are added in the figure below.

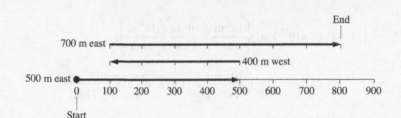

From the figure, note that the result of the sum of the three displacements puts the bee 800 m east of its starting point.
Assess: Knowing what a displacement is and how to add displacements, we are able to obtain the final position of the bee. Since the bee moved 1200 m to the east and 400 m to the west, it is reasonable that it would end up 800 m to the east of the starting point.

P1.9. Prepare: We are asked to rank in order three different speeds, so we simply compute each one according to Equation 1.1:

$$\text{speed} = \frac{\text{distance traveled in a given time interval}}{\text{time interval}}$$

Solve: (i) Toy

$$\frac{0.15 \text{ m}}{2.5 \text{ s}} = 0.060 \text{ m/s}$$

(ii) Ball

$$\frac{2.3 \text{ m}}{0.55 \text{ s}} = 4.2 \text{ m/s}$$

(iii) Bicycle

$$\frac{0.60 \text{ m}}{0.075 \text{ s}} = 8.0 \text{ m/s}$$

(iv) Cat

$$\frac{8.0 \text{ m}}{2.0 \text{ s}} = 4.0 \text{ m/s}$$

So the order from fastest to slowest is bicycle, ball, cat, and toy car.
Assess: We reported all answers to two significant figures as we should according to the significant figure rules. The result is probably what we would have guessed before solving the problem, although the cat and ball are close. These numbers all seem reasonable for the respective objects.

P1.13. Prepare: In this problem we are given $x_i = 2.1$ m and $x_f = 7.3$ m as well as $v = 0.35$ m/s and asked to solve for Δt.
Solve: We first solve for Δt in $v = \Delta x / \Delta t$ and then apply $\Delta x = x_f - x_i$.

$$\Delta t = \frac{\Delta x}{v} = \frac{x_f - x_i}{v} = \frac{7.3 \text{ m} - 2.1 \text{ m}}{0.35 \text{ m/s}} = \frac{5.2 \text{ m}}{0.35 \text{ m/s}} = 15 \text{ s}$$

Assess: 15 s seems like a long time for a ball to roll, but it is going fairly slowly, so the answer is reasonable.

P1.15. Prepare: We first collect the necessary conversion factors: 1 in = 2.54 cm; 1 cm = 10^{-2} m; 1 ft = 12 in; 39.37 in = 1 m; 1 mi = 1.609 km; 1 km = 10^3 m; 1 h = 3600 s.
Solve:

(a) $8.0 \text{ in} = 8.0 \text{ (in)} \left(\dfrac{2.54 \text{ cm}}{1 \text{ in}} \right) \left(\dfrac{10^{-2} \text{ m}}{1 \text{ cm}} \right) = 0.20 \text{ m}$

(b) $66 \text{ ft/s} = 66 \left(\dfrac{\text{ft}}{\text{s}} \right) \left(\dfrac{12 \text{ in}}{1 \text{ ft}} \right) \left(\dfrac{1 \text{ m}}{39.37 \text{ in}} \right) = 20 \text{ m/s}$

(c) $60 \text{ mph} = 60 \left(\dfrac{\text{mi}}{\text{h}} \right) \left(\dfrac{1.609 \text{ km}}{1 \text{ mi}} \right) \left(\dfrac{10^3 \text{ m}}{1 \text{ km}} \right) \left(\dfrac{1 \text{ h}}{3600 \text{ s}} \right) = 27 \text{ m/s}$

P1.19. Prepare: Review the rules for significant figures in Section 1.4 of the text, paying particular attention to the rules for addition (and subtraction) and multiplication (and division).

Solve: (a) $33.3 \times 25.4 = 846$

(b) $33.3 - 25.4 = 7.9$

(c) $\sqrt{33.3} = 5.77$

(d) $333.3 \div 25.4 = 13.1$

Assess: In part (a) the two numbers multiplied each have three significant figures and the answer has three significant figures. In part (b), even though each number has three significant figures, no information is significant past the tenths column. As a result, the answer is expressed only to the tenths column. In part (c), the number and the answer both have three significant figures. In part (d) the answer is expressed to three significant figures since this is the least number of significant figures in either of the two numbers in the problem.

P1.21. Prepare: Table 1.3 supplies the conversion factor we need: $1 \text{ ft} = 0.305 \text{ m}$.

Solve:

$$1250 \text{ ft} = 1250 \text{ ft} \left(\frac{0.305 \text{ m}}{1 \text{ ft}} \right) = 381 \text{ m} = 3.81 \times 10^2 \text{ m}$$

Assess: Since a meter is just bigger than a yard 381 m is about four football fields long. This seems to be right for the height of the Empire State Building.

P1.25. Prepare: In this problem we need to find the distance between Fort Collins and Greeley. We'll use the Pythagorean theorem.

Solve:

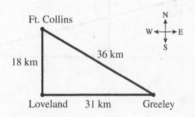

$$\sqrt{(18 \text{ km})^2 + (31 \text{ km})^2} = 36 \text{ km}$$

Assess: This is a reasonable distance between cities.

P1.27. Prepare: Since during part of the motion you are traveling north, and then west, the rules of vector addition can be used to determine the net displacement.

Solve: Since you've traveled 3 blocks north, and each block is 135 m long, the total distance traveled north is $(3)(135 \text{ m}) = 405 \text{ m}$. Two blocks west corresponds to $(2)(135 \text{ m}) = 270 \text{ m}$ (to three significant figures). The vectors form a right triangle. See the following vector diagram.

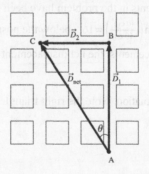

The length of the net displacement vector is $\overline{AC} = \sqrt{\overline{AB}^2 + \overline{BC}^2} = \sqrt{(270 \text{ m})^2 + (405 \text{ m})^2} = 487 \text{ m}$

Though it's not asked for in this problem, the angle is given by

$$\theta = \arctan\left(\frac{405 \text{ m}}{270 \text{ m}}\right) = 56.3°$$

So you have traveled 487 m 56.3° west of north.

Assess: For displacements in different directions you can use the law of vector addition to find the net displacement.

P1.31. Prepare: Knowing that the total trip consists of two displacements, we can add the two displacements to determine the total displacement and hence the distance of the goose from its original position. A quick sketch will help you visualize the two displacements and the total displacement.

Solve: The distance of the goose from its original position is the magnitude of the total displacement vector. This is determined as follows:

$$d = \sqrt{(32 \text{ km})^2 + (20 \text{ km})^2} = 38 \text{ km}$$

Assess: A quick look at your sketch shows that the total distance should be larger than the largest leg of the trip and this is the case.

P1.35. Prepare: In this problem we have three displacements to add using the laws of vector addition.
Solve: See the diagram below.

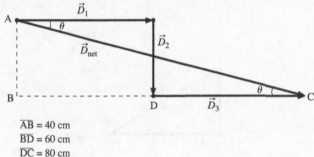

$$\overline{AB} = 40 \text{ cm}$$
$$\overline{BD} = 60 \text{ cm}$$
$$\overline{DC} = 80 \text{ cm}$$
$$\overline{BC} = 60 \text{ cm} + 80 \text{ cm} = 140 \text{ cm}$$

A convenient place to place the origin is the origin of the motion. We could add the first two vectors and then add the third vector on to that result. However, by looking at the diagram, the total displacement in the x direction is 60 cm + 80 cm = 140 cm to the right. The total displacement in the y direction is 40 cm downward. The net displacement will reflect these two displacements, so the result of the vector addition will be a vector pointing 140 cm to the right and 40 cm downward. Considering the right triangle ABC in the diagram, the magnitude of the displacement vector is then

$$\overline{AC} = \sqrt{\overline{AB}^2 + \overline{BC}^2} = \sqrt{(140 \text{ cm})^2 + (40 \text{ cm})^2} = 150 \text{ cm}$$

Where we have assumed that the measurements in the problem have been given to two significant figures.

The angle that the vector makes is $\theta = \arctan\left(\frac{40 \text{ cm}}{140 \text{ cm}}\right) = 16°$ below the positive x-axis.

Assess: When adding any number of displacements, the net displacement is always the vector between the initial and final point of the motion.

P1.37. Prepare: Draw a diagram of the situation. Note the right triangle.

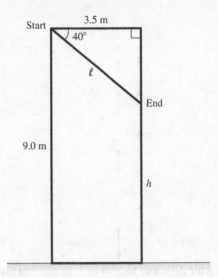

Solve: Use knowledge about right triangles from trigonometry.
(a) The length of the path is

$$\ell = \frac{3.5 \text{ m}}{\cos 40°} = 4.6 \text{ m}$$

(b) The height h above the ground is the vertical side of the triangle subtracted from 9.0 m.

$$h = 9.0 \text{ m} - (3.5 \text{ m})(\tan 40°) = 6.1 \text{ m}$$

Assess: Because of the angle we expect the squirrel to lose less height than the horizontal distance traveled.

P1.41. Prepare: Sam is moving along the x-axis. His speed changes from 60 mph to 30 mph during 3 s, which indicates slowing down (decreasing velocity vectors). Before braking, Sam drives at a constant speed.
Solve:

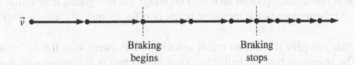

Braking
begins

Braking
stops

Assess: When the car is traveling at a constant speed (before and after braking) the position dots are uniformly spaced and the velocity vectors are constant in length. During the time the car is slowing the velocity vectors are decreasing in length and the position dots are getting closer.

P1.43. Prepare: Since the eland has a positive velocity but is slowing down, the velocity will decrease to zero and the spacing between the position dots will decrease. The velocity vector at each position on the way up has the same magnitude but opposite direction as the velocity at each position on the way down.

Solve:

Going up — | — Coming down

Assess: On the way up, the velocity vector decreases to zero as it should and the spacing between the position dots decreases as it should. The magnitude of the velocity vector at any position is the same on the way up as it is on the way down. This allows us to conclude that the figure is correct.

P1.47. Prepare: Knowing the dots represent the position of an object at equal time intervals and the vectors represent the velocity of the object at these times, we can construct a situation to match the motion diagram.
Solve: Reema passes 3rd Street doing 40 mph, slows steadily to the stop sign at 4th Street, stops for 1 s, then speeds up and reaches her original speed as she passes 5th Street. If the blocks are 50 m long, how long does it take Reema to drive from 3rd Street to 5th Street?
Assess: The statement that Reema slows to a stop in one block and regains her initial velocity in one block is consistent with the symmetry of the position dots and the velocity vectors about the stop position.

P1.49. Prepare: Knowing the dots represent the position of an object at equal time intervals and the vectors represent the velocity of the object at these times, we can construct a situation to match the motion diagram.
Solve: A bowling ball is at rest at the top of an incline. You nudge the ball giving it an initial velocity and causing it to roll down an incline. At the bottom of the incline it bounces off a sponge and travels back up the incline until it stops.
Assess: The statement that you give the ball an initial velocity is consistent with the fact that the start position dot has a velocity vector. The statement that the ball rolls down the incline is consistent with the fact that the dots are getting farther apart and the velocity vectors are increasing in length. The statement that the ball bounces off a sponge is consistent with the fact that ball does not bounce back to its original position.

P1.53. Prepare: Given a distance (32 mi) and a time (45 min) we are asked to compute the required speed. Since we want the answer in mph, a preliminary unit conversion will be helpful: 45 min = 0.75 h.
Solve:

$$\text{speed} = \frac{\text{distance traveled in a given time interval}}{\text{time interval}} = \frac{32 \text{ mi}}{0.75 \text{ h}} = 43 \text{ mph}$$

Assess: This is, of course, the *slowest* speed Alberta could drive and still arrive in time; any faster speed would simply get her there earlier.

P1.57. Prepare: Knowing that speed is distance divided by time, the distance is the speed multiplied by the time. 15 min = ¼ h.
Solve:
(a) The distance traveled during the ¼ hour is

$$\text{distance} = \text{speed} \times \text{time} = (100 \text{ km/h})(0.25 \text{ h}) = 25 \text{ km}$$

(b) Since the circumference of the track is 12.5 km, then the car goes completely around the track exactly twice in covering 25 km. Hence, the displacement from the initial position is 0 km.

(c) The speed of the car is

$$v = \left(100\frac{\text{km}}{\text{h}}\right)\left(\frac{1000\text{ m}}{1\text{ km}}\right)\left(\frac{1\text{ h}}{60\text{ min}}\right)\left(\frac{1\text{ min}}{60\text{ s}}\right) = 28\text{ m/s}$$

Assess: 28 m/s seems like a reasonable speed for a fast car.

P1.59. Prepare: Knowing the distance and time to travel that distance, we can determine the speed. Since we lack detailed information about the flight, and knowing that the flight is made by a bird, it is conceivable that its actual path darted back and forth and maybe even backtracked at times. If that is the case its actual distance of travel and hence average speed would be larger.

Solve: (a) The minimum average speed of the albatross may be determined by

$$v = \frac{\text{distance}}{\text{time}} = \left(\frac{1.2\times10^3\text{ km}}{1.4\text{ day}}\right)\left(\frac{1\text{ day}}{24\text{ hr}}\right)\left(\frac{0.621\text{ mi}}{1\text{ km}}\right) = 22\text{ mph}$$

(b) The average speed of the bird is 22 mph if it flies in a straight line between the end points. If the bird deviates from this line, the average speed will have to be greater than 22 mph in order to have the 1200 km displacement in 1.4 days.

Assess: You can ride a bike 10 to 15 mph and while you are riding your bike, birds easily fly past you. In light of that this is a reasonable answer.

P1.63. Prepare: Draw a diagram of the situation. Note the right triangle.

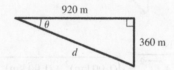

Solve: Use knowledge about right triangles from trigonometry.
(a) The angle of the path below the horizontal is

$$\theta = \arctan\left(\frac{360\text{ m}}{920\text{ m}}\right) = 21°$$

(b) The distance d covered is

$$d = \sqrt{(360\text{ m})^2 + (920\text{ m})^2} = 988\text{ m}$$

which should be reported as 990 m to two significant figures. We keep one more significant figure to use in the next step.
(c) The seal's speed is

$$v = \frac{\text{distance}}{\text{time}} = \left(\frac{988\text{ m}}{4.0\text{ min}}\right)\left(\frac{1\text{ min}}{60\text{ s}}\right) = 4.1\text{ m/s}$$

Assess: Because the seal descended less than the horizontal distance we expect the angle to be less than 45 degrees.

P1.65. Prepare: Draw a diagram of the situation. Note the right triangle.

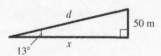

Solve: Use knowledge about right triangles from trigonometry.
(a) The horizontal distance x is

$$x = \frac{50\text{ m}}{\tan 13°} = 216.6\text{ m}$$

This should be reported as 220 m to two significant figures.

(b) The distance d covered is

$$d = \sqrt{(50 \text{ m})^2 + (216.6 \text{ m})^2} = 222.3 \text{ m}$$

which should be reported as 220 m to two significant figures. We keep two more significant figures to use in the next step.

(c) The time it takes the shark is

$$\text{time} = \frac{\text{distance}}{\text{speed}} = \left(\frac{222.3 \text{ m}}{0.85 \text{ m/s}} \right) = 260 \text{ s}$$

Assess: 260 s is just over 4 min, which is impressive but reasonable.

P1.67. Prepare: Since during part of the motion John is traveling north and then turns east, the rules of vector addition will be used to determine the net displacement.
Solve: (a) The vectors form a right triangle. See the following vector diagram.

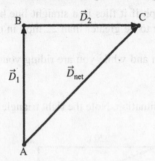

The length of the net displacement vector is

$$\overline{AC} = \sqrt{\overline{AB}^2 + \overline{BC}^2} = \sqrt{(1.00 \text{ km})^2 + (1.00 \text{ km})^2} = 1.41 \text{ km}$$

(b) Jane walks along John's net displacement vector, so she only travels 1.41 km, while John travels a total distance of 2.00 km. Since he travels at 1.50 m/s during the entire stroll, the time John takes to get to his destination is

$$\Delta t_{\text{John}} = \frac{2000 \text{ m}}{1.50 \text{ m/s}} = 1.33 \times 10^3 \text{ s}$$

For Jane to walk 1.41 km in this time, her velocity would need to be

$$v_{\text{Jane}} = \frac{1410 \text{ m}}{1.33 \times 10^3 \text{ s}} = 1.06 \text{ m/s}$$

Assess: Jane must walk slower than John to walk the shorter distance in the same time, so the answer makes sense. For displacements in different directions you must use the law of vector addition.

2

MOTION IN ONE DIMENSION

Q2.3. Reason: Where the rings are far apart the tree is growing rapidly. It appears that the rings are quite far apart near the center (the origin of the graph), then get closer together, then farther apart again.

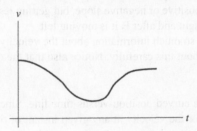

Assess: After drawing velocity-versus-time graphs (as well as others), stop and think if it matches the physical situation, especially by checking end points, maximum values, places where the slope is zero, etc. This one passes those tests.

Q2.5. Reason: Let $t_0 = 0$ be when you pass the origin. The other car will pass the origin at a later time t_1 and passes you at time t_2.

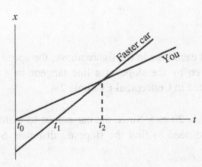

Assess: The slope of the position graph is the velocity, and the slope for the faster car is steeper.

Q2.7. Reason: A predator capable of running at a great speed while not being capable of large accelerations could overtake slower prey that were capable of large accelerations, given enough time. However, it may not be as effective as surprising and grabbing prey that are capable of higher acceleration. For example, prey could escape if the safety of a burrow were nearby. If a predator were capable of larger accelerations than its prey, while being slower in speed than the prey, it would have a greater chance of surprising and grabbing prey, quickly, though prey might outrun it if given enough warning.

Assess: Consider the horse-man race discussed in the text.

Q2.9. Reason: (a) Once the rock leaves the thrower's hand, it is in free fall. While in free fall, the acceleration of the rock is exactly the acceleration due to gravity, which has a magnitude g and is downward. The fact that the rock was thrown and not simply dropped means that the rock has an initial velocity when it leaves the thrower's hand. This does not affect the acceleration of gravity, which does not depend on how the rock was thrown.

(b) Just before the rock hits the water, it is still in free fall. Its acceleration remains the acceleration of gravity. Its velocity has increased due to gravity, but acceleration due to gravity is independent of velocity.

Assess: No matter what the velocity of an object is, the acceleration due to gravity always has magnitude g and is always straight downward.

Q2.13. Reason: (a) D. The steepness of the tangent line is greatest at D.

(b) C, D, E. Motion to the left is indicated by a decreasing segment on the graph.

(c) C. The speed corresponds to the steepness of the tangent line, so the question can be re-cast as "Where is the tangent line getting steeper (either positive or negative slope, but getting steeper)?" The slope at B is zero and is greatest at D, so it must be getting steeper at C.

(d) A, E. The speed corresponds to the steepness of the tangent line, so the question can be re-cast as "Where is the tangent line getting less steep (either positive or negative slope, but getting less steep)?"

(e) B. Before B the object is moving right and after B it is moving left.

Assess: It is amazing that we can get so much information about the velocity (and even about the acceleration) from a position-versus-time graph. Think about this carefully. Notice also that the object is at rest (to the left of the origin) at point F.

Q2.15. Reason: This graph shows a curved position-versus-time line. Since the graph is curved the motion is *not* uniform. The instantaneous velocity, or the velocity at any given instant of time, is the slope of a line tangent to the graph at that point in time. Consider the graph below, where tangents have been drawn at each labeled time.

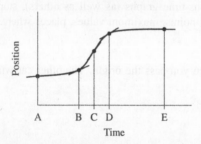

Comparing the slope of the tangents at each time in the figure above, the speed of the car is greatest at time C.

Assess: Instantaneous velocity is given by the slope of a line tangent to a position-versus-time curve at a given instant of time. This is also demonstrated in Conceptual Example 2.4.

Q2.19. Reason: The initial velocity is 20 m/s. Since the car comes to a stop, the final velocity is 0 m/s. We are given the acceleration of the car, and need to find the stopping distance. See the pictorial representation, which includes a list of values below.

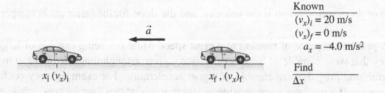

Known
$(v_x)_i = 20$ m/s
$(v_x)_f = 0$ m/s
$a_x = -4.0$ m/s^2

Find
Δx

An equation that relates acceleration, initial velocity, final velocity, and distance is Equation 2.13.

$$(v_x)_f^2 = (v_x)_i^2 + 2a_x\Delta x$$

Solving for Δx,

$$\Delta x = \frac{(v_x)_f^2 - (v_x)_i^2}{2a_x} = \frac{(0 \text{ m/s})^2 - (20 \text{ m/s})^2}{2(-4.0 \text{ m/s}^2)} = 50 \text{ m}$$

The correct choice is D.

Assess: We are given initial and final velocities and acceleration. We are asked to find a displacement, so Equation 2.13 is an appropriate equation to use.

Q2.21. Reason: The slope of the tangent to the velocity-versus-time graph gives the acceleration of each car. At time $t = 0$ s the slope of the tangent to Andy's velocity-versus-time graph is very small. The slope of the tangent to the graph at the same time for Carl is larger. However, the slope of the tangent in Betty's case is the largest of the three. So Betty had the greatest acceleration at $t = 0$ s. See the figure below.

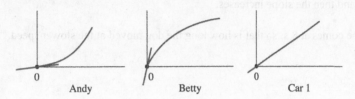

The correct choice is B.

Assess: Acceleration is given by the slope of the tangent to the curve in a velocity-versus-time graph at a given time.

Q2.25. Reason: This can be solved with simple ratios. Since $\vec{a} = \dfrac{\Delta \vec{v}}{\Delta t}$ and the a stays the same, it would take twice as long to change $\vec{v}$ twice as much.

The answer is B.

Assess: This result can be checked by actually computing the acceleration and plugging it back into the equation for the second case, but ratios are slicker and quicker.

Problems

P2.1. Prepare: The car is traveling to the left toward the origin, so its position decreases with increase in time.

Solve: (a)

Time t (s)	Position x (m)
0	1200
1	975
2	825
3	750
4	700
5	650
6	600
7	500
8	300
9	0

(b)

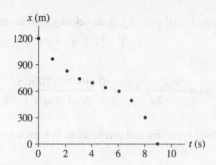

Assess: A car's motion traveling down a street can be represented at least three ways: a motion diagram, position-versus-time data presented in a table (part **(a)**), and a position-versus-time graph (part **(b)**).

P2.3. Prepare: The slope of the position graph is the velocity graph. The position graph has a shallow (negative) slope for the first 8 s, and then the slope increases.
Solve:
(a) The change in slope comes at 8 s, so that is how long the dog moved at the slower speed.
(b)

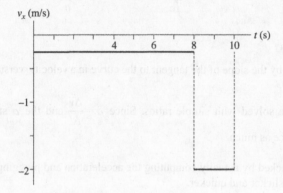

Assess: We expect the sneaking up phase to be longer than the spring phase, so this looks like a realistic situation.

P2.9. Prepare: A visual overview of Alan's and Beth's motion that includes a pictorial representation, a motion diagram, and a list of values is shown below. Our strategy is to calculate and compare Alan's and Beth's time of travel from Los Angeles to San Francisco.

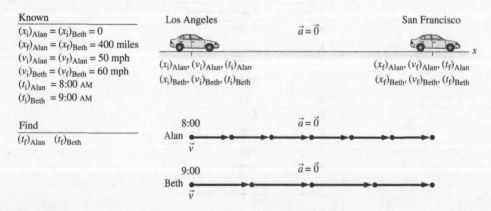

Solve: Beth and Alan are moving at a constant speed, so we can calculate the time of arrival as follows:

$$v = \frac{\Delta x}{\Delta t} = \frac{x_f - x_i}{t_f - t_i} \Rightarrow t_f = t_i + \frac{x_f - x_i}{v}$$

Using the known values identified in the pictorial representation, we find

$$(t_f)_{Alan} = (t_i)_{Alan} + \frac{(x_f)_{Alan} - (x_i)_{Alan}}{v} = 8:00 \text{ AM} + \frac{400 \text{ mile}}{50 \text{ miles/hour}} = 8:00 \text{ AM} + 8 \text{ hr} = 4:00 \text{ PM}$$

$$(t_f)_{Beth} = (t_i)_{Beth} + \frac{(x_f)_{Beth} - (x_i)_{Beth}}{v} = 9:00 \text{ AM} + \frac{400 \text{ mile}}{60 \text{ miles/hour}} = 9:00 \text{ AM} + 6.67 \text{ hr} = 3:40 \text{ PM}$$

(a) Beth arrives first.
(b) Beth has to wait 20 minutes for Alan.
Assess: Times of the order of 7 or 8 hours are reasonable in the present problem.

P2.11. Prepare: Since each runner is running at a steady pace, they both are traveling with a constant speed. Each must travel the same distance to finish the race. We assume they are traveling uniformly. We can calculate the time it takes each runner to finish using Equation 2.1.
Solve: The first runner finishes in

$$\Delta t_1 = \frac{\Delta x}{(v_x)_1} = \frac{5.00 \text{ km}}{12.0 \text{ km/h}} = 0.417 \text{ h}$$

Converting to minutes, this is $(0.417 \text{ h})\left(\frac{60 \text{ min}}{1 \text{ h}}\right) = 25.0 \text{ min}$

For the second runner

$$\Delta t_2 = \frac{\Delta x}{(v_x)_2} = \frac{5.00 \text{ km}}{14.5 \text{ km/h}} = 0.345 \text{ h}$$

Converting to seconds, this is

$$(0.345 \text{ h})\left(\frac{60 \text{ min}}{1 \text{ h}}\right) = 20.7 \text{ min}$$

The time the second runner waits is 25.0 min – 20.7 min = 4.3 min
Assess: For uniform motion, velocity is given by Equation 2.1.

P2.13. Prepare: Assume v_x is constant so the ratio $\frac{\Delta x}{\Delta t}$ is also constant.

Solve:
(a)

$$\frac{30 \text{ m}}{3.0 \text{ s}} = \frac{\Delta x}{1.5 \text{ s}} \Rightarrow \Delta x = 1.5 \text{ s}\left(\frac{30 \text{ m}}{3.0 \text{ s}}\right) = 15 \text{ m}$$

(b)

$$\frac{30 \text{ m}}{3.0 \text{ s}} = \frac{\Delta x}{9.0 \text{ s}} \Rightarrow \Delta x = 9.0 \text{ s}\left(\frac{30 \text{ m}}{3.0 \text{ s}}\right) = 90 \text{ m}$$

Assess: Setting up the ratio allows us to easily solve for the distance traveled in any given time.

P2.17. Prepare: Since displacement is equal to the area under the velocity graph between t_i and t_f, we can find the car's final position from its initial position and the area.
Solve: (a) Using the equation $x_f = x_i$ + area of the velocity graph between t_i and t_f,

$$x_{2s} = 10 \text{ m} + \text{area of trapezoid between 0 s and 2 s}$$

$$= 10 \text{ m} + \frac{1}{2}(12 \text{ m/s} + 4 \text{ m/s})(2 \text{ s}) = 26 \text{ m}$$

$$x_{3\,s} = 10 \text{ m} + \text{area of triangle between 0 s and 3 s}$$

$$= 10 \text{ m} + \frac{1}{2}(12 \text{ m/s})(3 \text{ s}) = 28 \text{ m}$$

$$x_{4\,s} = x_{3\,s} + \text{area between 3 s and 4 s}$$

$$= 28 \text{ m} + \frac{1}{2}(-4 \text{ m/s})(1 \text{ s}) = 26 \text{ m}$$

(b) The car reverses direction at $t = 3$ s, because its velocity becomes negative.

Assess: The car starts at $x_i = 10$ m at $t_i = 0$. Its velocity decreases as time increases, is zero at $t = 3$ s, and then becomes negative. The slope of the velocity-versus-time graph is negative which means the car's acceleration is negative and a constant. From the acceleration thus obtained and given velocities on the graph, we can also use kinematic equations to find the car's position at various times.

P2.19. Prepare: The graph in Figure P2.19 shows distinct slopes in the time intervals: $0 - 2$ s and 2 s $- 4$ s. We can thus obtain the acceleration values from this graph using $a_x = \Delta v_x/\Delta t$. A linear decrease in velocity from $t = 0$ s to $t = 2$ s implies a constant negative acceleration. On the other hand, a constant velocity between $t = 2$ s and $t = 4$ s means zero acceleration.

Solve: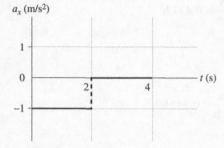

P2.23. Prepare: From a velocity-versus-time graph we find the acceleration by computing the slope. We will compute the slope of each straight-line segment in the graph.

$$a_x = \frac{(v_x)_f - (v_x)_i}{t_f - t_i}$$

The trickiest part is reading the values off of the graph.
Solve: (a)

$$a_x = \frac{5.5 \text{ m/s} - 0.0 \text{ m/s}}{0.9 \text{ s} - 0.0 \text{ s}} = 6.1 \text{ m/s}^2$$

(b)

$$a_x = \frac{9.3 \text{ m/s} - 5.5 \text{ m/s}}{2.4 \text{ s} - 0.9 \text{ s}} = 2.5 \text{ m/s}^2$$

(c)

$$a_x = \frac{10.9 \text{ m/s} - 9.3 \text{ m/s}}{3.5 \text{ s} - 2.4 \text{ s}} = 1.5 \text{ m/s}^2$$

Assess: This graph is difficult to read to more than one significant figure. I did my best to read a second significant figure but there is some estimation in the second significant figure.
It takes Carl Lewis almost 10 s to run 100 m, so this graph covers only the first third of the race. Were the graph to continue, the slope would continue to decrease until the slope is zero as he reaches his (fastest) cruising speed.
Also, if the graph were continued out to the end of the race, the area under the curve should total 100 m.

P2.25. Prepare: We can calculate acceleration from Equation 2.8:
Solve: For the gazelle:

$$(a_x) = \left(\frac{\Delta v_x}{\Delta t}\right) = \frac{13 \text{ m/s}}{3.0 \text{ s}} = 4.3 \text{ m/s}^2$$

For the lion:

$$(a_x) = \left(\frac{\Delta v_x}{\Delta t}\right) = \frac{9.5 \text{ m/s}}{1.0 \text{ s}} = 9.5 \text{ m/s}^2$$

For the trout:

$$(a_x) = \left(\frac{\Delta v_x}{\Delta t}\right) = \frac{2.8 \text{ m/s}}{0.12 \text{ s}} = 23 \text{ m/s}^2$$

The trout is the animal with the largest acceleration.
Assess: A lion would have an easier time snatching a gazelle than a trout.

P2.27. Prepare: First, we will convert units:

$$60 \frac{\text{miles}}{\text{hour}} \times \frac{1 \text{ hour}}{3600 \text{ s}} \times \frac{1609 \text{ m}}{1 \text{ mile}} = 26.8 \text{ m/s}$$

We also note that $g = 9.8 \text{ m/s}^2$. Because the car has constant acceleration, we can use kinematic equations.
Solve: **(a)** For initial velocity $v_i = 0$, final velocity $v_f = 26.8 \text{ m/s}$, and $\Delta t = 10$ s, we can find the acceleration using

$$v_f = v_i + a\Delta t \Rightarrow a = \frac{v_f - v_i}{\Delta t} = \frac{(26.8 \text{ m/s} - 0 \text{ m/s})}{10 \text{ s}} = 2.68 \text{ m/s}^2 \approx 2.7 \text{ m/s}^2$$

(b) The fraction is $a/g = 2.68/9.8 = 0.273$. So a is 27% of g, or 0.27 g.
(c) The displacement is calculated as follows:

$$x_f - x_i = v_i\Delta t + \frac{1}{2}a(\Delta t)^2 = \frac{1}{2}a(\Delta t)^2 = 134 \text{ m} = 440 \text{ feet}$$

Assess: A little over tenth of a mile displacement in 10 s is physically reasonable.

P2.31. Prepare: Because the skier slows steadily, her deceleration is a constant during the glide and we can use the kinematic equations of motion under constant acceleration.
Solve: Since we know the skier's initial and final speeds and the width of the patch over which she decelerates, we will use

$$v_f^2 = v_i^2 + 2a(x_f - x_i)$$

$$\Rightarrow a = \frac{v_f^2 - v_i^2}{2(x_f - x_i)} = \frac{(6.0 \text{ m/s})^2 - (8.0 \text{ m/s})^2}{2(5.0 \text{ m})} = -2.8 \text{ m/s}^2$$

The magnitude of this acceleration is 2.8 m/s^2.
Assess: A deceleration of 2.8 m/s^2 or 6.3 mph/s is reasonable.

P2.33. Prepare: Because the car slows steadily, the deceleration is a constant and we can use the kinematic equations of motion under constant acceleration.
Solve: Since we know the car's initial and final speeds and the width of the patch over which she decelerates, we will use

$$v_f^2 = v_i^2 + 2a(x_f - x_i)$$

$$\Rightarrow a = \frac{v_f^2 - v_i^2}{2(x_f - x_i)} = \frac{(0 \text{ m/s})^2 - (90 \text{ m/s})^2}{2(110 \text{ m})} = -37 \text{ m/s}^2$$

The magnitude of this acceleration is 37 m/s^2.
Assess: A deceleration of 37 m/s^2 is impressive; it is almost 4 gs.

P2.39.   Prepare: Use kinematic equations for constant acceleration. Call the point where the motorcycle started the origin.
Solve:
(a)

$$a = \frac{\Delta v}{\Delta t} \Rightarrow \Delta t = \frac{\Delta v}{a} = \frac{80 \text{ km/h}}{8.0 \text{ m/s}^2}\left(\frac{1 \text{ h}}{3600 \text{ s}}\right)\left(\frac{1000 \text{ m}}{1 \text{ km}}\right) = 2.78 \text{ s} \approx 2.8 \text{ s}$$

(b) Compute the distance traveled in 10 s for each vehicle.

For the car: $\Delta x = v\Delta t = (80 \text{ km/h})(2.78 \text{ s})\left(\frac{1 \text{ h}}{3600 \text{ s}}\right)\left(\frac{1000 \text{ m}}{1 \text{ km}}\right) = 61.7 \text{ m}$

For the motorcycle: $\Delta x = \frac{1}{2}a(\Delta t)^2 = \frac{1}{2}(8.0 \text{ m/s}^2)(2.78 \text{ s})^2 = 30.7 \text{ m}$

The difference is the distance between the motorcycle and the car at that time. 61.7 m − 30.7 m = 31 m
Assess: The motorcycle will never catch up if it never exceeds the speed of the car.

P2.41.   Prepare: We will use the equation for constant acceleration to find out how far the sprinter travels during the acceleration phase. Use Equation 2.11 to find the acceleration.

$$v_x = a_x t_1 \qquad \text{where } v_0 = 0 \text{ and } t_0 = 0$$

$$a_x = \frac{v_x}{t_1} = \frac{11.2 \text{ m/s}}{2.14 \text{ s}} = 5.23 \text{ m/s}^2$$

Solve: The distance traveled during the acceleration phase will be

$$\Delta x = \frac{1}{2}a_x(\Delta t)^2$$

$$= \frac{1}{2}(5.23 \text{ m/s}^2)(2.14 \text{ s})^2$$

$$= 12.0 \text{ m}$$

The distance left to go at constant velocity is 100 m − 12.0 m = 88.0 m. The time this takes at the top speed of 11.2 m/s is

$$\Delta t = \frac{\Delta x}{v_x} = \frac{88.0 \text{ m}}{11.2 \text{ m/s}} = 7.86 \text{ s}$$

The total time is 2.14 s + 7.86 s = 10.0 s.
Assess: This is indeed about the time it takes a world-class sprinter to run 100 m (the world record is a bit under 9.8 s).
Compare the answer to this problem with the accelerations given in Problem 2.23 for Carl Lewis.

P2.43.   Prepare: The bill must drop its own length. Assume it is in free fall.
Solve:

$$\Delta y = \frac{1}{2}g(\Delta t)^2 \Rightarrow \Delta t = \sqrt{\frac{2\Delta y}{g}} = \sqrt{\frac{2(0.16 \text{ m})}{9.8 \text{ m/s}^2}} = 0.18 \text{ s}$$

Assess: This is less than the typical 0.25 s reaction time, so most people miss the bill.

P2.45.   Prepare: Use kinematic equations for constant acceleration. Assume the gannet is in free fall during the dive.
Solve:

$$(v_y)_f^2 = (v_y)_i^2 + 2g\Delta y \Rightarrow \Delta y = \frac{(v_y)_f^2}{2g} = \frac{(32 \text{ m/s})^2}{2(9.8 \text{ m/s}^2)} = 52 \text{ m}$$

Assess: 52 meters seems a reasonable height from which to begin the dive.

P2.49. Prepare: Since the villain is hanging on to the ladder as the helicopter is ascending, he and the briefcase are moving with the same upward velocity as the helicopter. We can calculate the initial velocity of the briefcase, which is equal to the upward velocity of the helicopter. See the following figure.

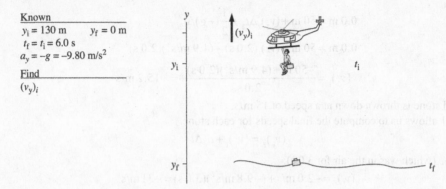

Known
$y_i = 130$ m $y_f = 0$ m
$t_f = t_i = 6.0$ s
$a_y = -g = -9.80$ m/s^2

Find
$(v_y)_i$

Solve: We can use Equation 2.12 here. We know the time it takes the briefcase to fall, its acceleration, and the distance it falls. Solving for $(v_y)_i \Delta t$,

$$(v_y)_i \Delta t = (y_f - y_i) - \frac{1}{2}(a_y)\Delta t^2 = -130 \text{ m} - \left[\frac{1}{2}(-9.80 \text{ m/s}^2)(6.0 \text{ s})^2 \right] = 46 \text{ m}$$

Dividing by Δt to solve for $(v_y)_i$,

$$(v_x)_i = \frac{46 \text{ m}}{6.0 \text{ s}} = 7.7 \text{ m/s}$$

Assess: Note the placement of negative signs in the calculation. The initial velocity is positive, as expected for a helicopter ascending.

P2.51. Prepare: There are several steps in this problem, so first draw a picture and, like the examples in the book, list the known quantities and what we need to find.
Call the pool of water the origin and call $t = 0$ s when the first stone is released. We will assume both stones are in free fall after they leave the climber's hand, so $a_y = -g$. Let a subscript 1 refer to the first stone and a 2 refer to the second.

Known	Find
$(y_1)_i = 50$ m	$(t_2)_f$ or t_f
$(y_2)_i = 50$ m	$(v_2)_i$
$(y_1)_f = 0.0$ m	$(v_1)_f$
$(y_2)_f = 0.0$ m	$(v_2)_f$
$(y_1)_i = -2.0$ m/s	
$(t_2)_f = (t_1)_f$; simply call this t_f	
$(t_2)_i = 1.0$ s	

Solve: (a) Using $(t_1)_i = 0$

$$(y_1)_f = (y_1)_i + (v_1)_i \Delta t + \frac{1}{2} a_y \Delta t^2$$

$$0.0 \text{ m} = 50 \text{ m} + (-2 \text{ m/s})t_f + \frac{1}{2}(-g)t_f^2$$

$$0.0 \text{ m} = 50 \text{ m} - (2 \text{ m/s})t_f - (4.9 \text{ m/s}^2)t_f^2$$

Solving this quadratic equation gives two values for t_f: 3.0 s and −3.4 s, the second of which (being negative) is outside the scope of this problem.
Both stones hit the water at the same time, and it is at $t = 3.0$ s, or 3.0 s after the first stone is released.

(b) For the second stone $\Delta t_2 = t_f - (t_2)_i = 3.0 \text{ s} - 1.0 \text{ s} = 2.0 \text{ s}$. We solve now for $(v_2)_i$.

$$(y_2)_f = (y_2)_i + (v_2)_i \Delta t + \frac{1}{2} a_y \Delta t^2$$

$$0.0 \text{ m} = 50 \text{ m} + (v_2)_i \Delta t_2 + \frac{1}{2}(-g)\Delta t_2^2$$

$$0.0 \text{ m} = 50 \text{ m} + (v_2)_i (2.0 \text{ s}) - (4.9 \text{ m/s}^2)(2.0 \text{ s})^2$$

$$(v_2)_i = \frac{-50 \text{ m} + (4.9 \text{ m/s}^2)(2.0 \text{ s})^2}{2.0 \text{ s}} = -15.2 \text{ m/s}$$

Thus, the second stone is thrown down at a speed of 15 m/s.

(c) Equation 2.11 allows us to compute the final speeds for each stone.

$$(v_y)_f = (v_y)_i + a_y \Delta t$$

For the first stone (which was in the air for 3.0 s):

$$(v_1)_f = -2.0 \text{ m/s} + (-9.8 \text{ m/s}^2)(3.0 \text{ s}) = -31 \text{ m/s}$$

The speed is the magnitude of this velocity, or 31 m/s.
For the second stone (which was in the air for 2.0 s):

$$(v_2)_f = -15.2 \text{ m/s} + (-9.8 \text{ m/s}^2)(2.0 \text{ s}) = -35 \text{ m/s}$$

The speed is the magnitude of this velocity, or 35 m/s.

Assess: The units check out in each of the previous equations. The answers seem reasonable. A stone dropped from rest takes 3.2 s to fall 50 m; this is comparable to the first stone, which was able to fall the 50 m in only 3.0 s because it started with an initial velocity of −2.0 m/s. So we are in the right ballpark. And the second stone would have to be thrown much faster to catch up (because the first stone is accelerating).

P2.53. Prepare: Assume the truck driver is traveling with constant velocity during each segment of his trip.
Solve: Since the driver usually takes 8 hours to travel 440 miles, his usual velocity is

$$v_{\text{usual } x} = \frac{\Delta x}{\Delta t_{\text{usual}}} = \frac{440 \text{ mi}}{8 \text{ h}} = 55 \text{ mph}$$

However, during this trip he was driving slower for the first 120 miles. Usually he would be at the 120 mile point in

$$\Delta t_{\text{usual at 120 mi}} = \frac{\Delta x}{v_{\text{usual at 120 mi } x}} = \frac{120 \text{ mi}}{55 \text{ mph}} = 2.18 \text{ h}$$

He is 15 minutes, or 0.25 hr late. So the time he's taken to get 120 mi is 2.18 hr + 0.25 hr = 2.43 hr. He wants to complete the entire trip in the usual 8 hours, so he only has 8 hr − 2.43 hr = 5.57 hr left to complete 440 mi − 120 mi = 320 mi. So he needs to increase his velocity to

$$v_{\text{to catch up } x} = \frac{\Delta x}{\Delta t_{\text{to catch up}}} = \frac{320 \text{ mi}}{5.57 \text{ h}} = 57 \text{ mph}$$

where additional significant figures were kept in the intermediate calculations.
Assess: This result makes sense. He is only 15 minutes late.

P2.59. Prepare: After appropriate unit conversions, we'll see how far the spacecraft goes during the acceleration phase and what speed it achieves and then how long it would take to go the remaining distance at that speed.

$$0.50 \text{ y} = 1.578 \times 10^7 \text{ s}$$

Solve: Because $(v_x)_i = 0$ m/s and $x_i = 0$ m

$$x_f = \frac{1}{2} a_x (\Delta t)^2 = \frac{1}{2}(9.8 \text{ m/s}^2)(1.578 \times 10^7 \text{ s})^2 = 1.220 \times 10^{15} \text{ m}$$

which is not a very large fraction of the whole distance. The spacecraft must still go $4.1 \times 10^{16} \text{ m} - 1.220 \times 10^{15} \text{ m}$
$= 3.98 \times 10^{16} \text{ m}$ at the achieved speed.

The speed is

$$\Delta v_x = a_x \Delta t = (9.8 \text{ m/s}^2)(1.578 \times 10^7 \text{ s}) = 1.55 \times 10^8 \text{ m/s}$$

which is half the speed of light. The time taken to go the remaining distance at that speed is

$$\Delta t = \frac{\Delta x}{v_x} = \frac{3.98 \times 10^{16} \text{ m}}{1.55 \times 10^8 \text{ m/s}} = 2.57 \times 10^8 \text{ s} = 8.15 \text{ y}$$

Now the total time needed is the sum of the time for the acceleration phase and the time for the constant velocity phase.

$$\Delta t = 0.50 \text{ y} + 8.15 \text{ y} = 8.7 \text{ y}$$

Assess: It is now easy to see why travel to other stars will be so difficult. We even made some overly generous assumptions and ignored relativistic effects.

P2.61. Prepare: Remember that in estimation problems different people may make slightly different estimates. That is OK as long as they end up with reasonable answers that are the same order-of-magnitude.
By assuming the acceleration to be constant we can use

$$x_f = \frac{1}{2} a_x (\Delta t)^2$$

Solve: (a) I guessed about 1.0 cm; this was verified with a ruler and mirror.
(b) We are given a closing time of 0.024 s, so we can compute the acceleration from rearranging the kinematic equations.

$$a_x = \frac{2x_f}{(\Delta t)^2} = \frac{2(1.0 \text{ cm})}{(0.024 \text{ s})^2} \left(\frac{1 \text{ m}}{100 \text{ cm}} \right) = 35 \text{ m/s}^2$$

(c) Since we know the Δt and the a and $v_i = 0.0$ m/s, we can compute the final speed from Equation 2.11:

$$v_f = a \Delta t = (35 \text{ m/s}^2)(0.024 \text{ s}) = 0.84 \text{ m/s}$$

Assess: The uncertainty in our estimates might or might not barely justify two significant figures.
The final speed is reasonable; if we had arrived at an answer 10 times bigger or 10 times smaller we would probably go back and check our work. The lower lid gets smacked at this speed up to 15 times per minute!

P2.65. Prepare: A visual overview of the ball's motion that includes a pictorial representation, a motion diagram, and a list of values is shown below. We label the ball's motion along the y-axis. As soon as the ball leaves the student's hand, it is falling freely and thus kinematic equations hold. The ball's acceleration is equal to the acceleration due to gravity that always acts vertically downward toward the center of the earth. The initial position of the ball is at the origin where $y_i = 0$, but the final position is below the origin at $y_f = -2.0$ m. Recall sign conventions, which tell us that v_i is positive and a is negative.

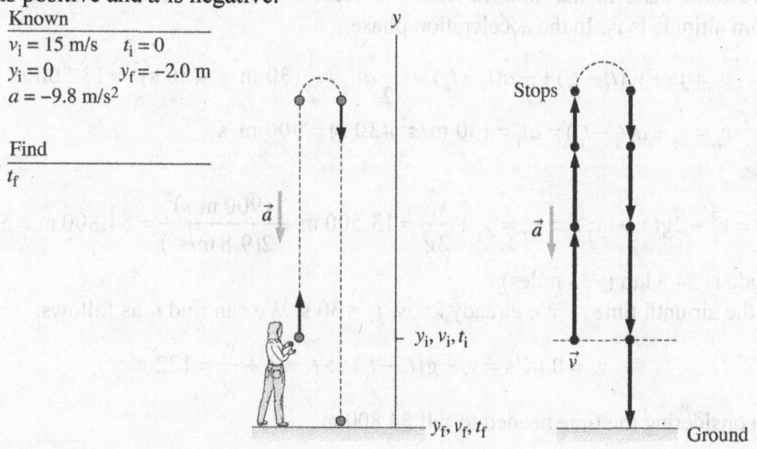

Solve: With all the known information, it is clear that we must use

$$y_f = y_i + v_i \Delta t + \frac{1}{2} a \Delta t^2$$

Substituting the known values

$$-2 \text{ m} = 0 \text{ m} + (15 \text{ m/s})t_f + (1/2)(-9.8 \text{ m/s}^2)t_f^2$$

The solution of this quadratic equation gives $t_f = 3.2$ s. The other root of this equation yields a negative value for t_f, which is not physical for this problem.

Assess: A time of 3.2 s is reasonable.

P2.67. Prepare: A visual overview of the rocket's motion that includes a pictorial representation, a motion diagram, and a list of values is shown below. We represent the rocket's motion along the y-axis. The rocket accelerates upward for 30 s, but as soon as the rocket runs out of fuel, it falls freely. The kinematic equations hold separately before as well as after the rocket runs out of fuel because accelerations for both situations are constant, 30.0 m/s² for the former and 9.8 m/s² for the latter. Also, note that $a_0 = +30.0$ m/s² is vertically upward, but $a_1 = a_2 = -9.8$ m/s² acts vertically downward. This is a three-part problem. For the first accelerating phase, the initial position of the rocket is at the origin where $y_0 = 0$, but the position when fuel runs out is at y_1. Recall sign conventions, which tell us that v_0 is positive. From the given information, we can find v_1. For the second part of the problem, v_1 is positive as the rocket is moving upward, v_2 is zero as it reaches the maximum altitude, and a_1 is negative. This helps us find y_2. The third part involves finding t_2 and t_3, which can be obtained using kinematics.

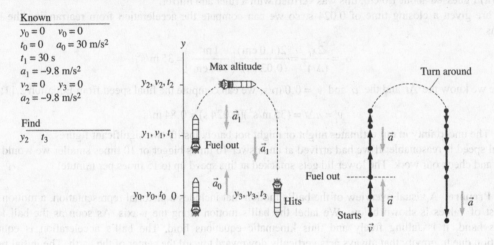

Solve: (a) There are three parts to the motion. Both the second and third parts of the motion are free fall, with $a = -g$. The maximum altitude is y_2. In the acceleration phase

$$y_1 = y_0 + v_0(t_1 - t_0) + \frac{1}{2}a(t_1 - t_0)^2 = \frac{1}{2}at_1^2 = \frac{1}{2}(30 \text{ m/s}^2)(30 \text{ s})^2 = 13{,}500 \text{ m}$$

$$v_1 = v_0 + a(t_1 - t_0) = at_1 = (30 \text{ m/s}^2)(30 \text{ s}) = 900 \text{ m/s}$$

In the coasting phase,

$$v_2^2 = 0 = v_1^2 - 2g(y_2 - y_1) \Rightarrow y_2 = y_1 + \frac{v_1^2}{2g} = 13{,}500 \text{ m} + \frac{(900 \text{ m/s})^2}{2(9.8 \text{ m/s}^2)} = 54{,}800 \text{ m} = 54.8 \text{ km}$$

The maximum altitude is 54.8 km ($\approx$33 miles).

(b) The rocket is in the air until time t_3. We already know $t_1 = 30$ s. We can find t_2 as follows:

$$v_2 = 0 \text{ m/s} = v_1 - g(t_2 - t_1) \Rightarrow t_2 = t_1 + \frac{v_1}{g} = 122 \text{ s}$$

Then t_3 is found by considering the time needed to fall 54,800 m:

$$y_3 = 0 \text{ m} = y_2 + v_2(t_3 - t_2) - \frac{1}{2}g(t_3 - t_2)^2 = y_2 - \frac{1}{2}g(t_3 - t_2)^2 \Rightarrow t_3 = t_2 + \sqrt{\frac{2y_2}{g}} = 230 \text{ s}$$

(c) The velocity increases linearly, with a slope of 30 (m/s)/s, for 30 s to a maximum speed of 900 m/s. It then begins to decrease linearly with a slope of -9.8 (m/s)/s. The velocity passes through zero (the turning point at y_2) at $t_2 = 122$ s. The impact velocity at $t_3 = 230$ s is calculated to be $v_3 = v_2 - g(t_3 - t_2) = -1000$ m/s.

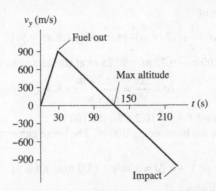

Assess: In reality, friction due to air resistance would prevent the rocket from reaching such high speeds as it falls, and the acceleration upward would not be constant because the mass changes as the fuel is burned, but that is a more complicated problem.

P2.69. Prepare: A visual overview of car's motion that includes a pictorial representation, a motion diagram, and a list of values is shown below. We label the car's motion along the x-axis. This is a three-part problem. First the car accelerates, then it moves with a constant speed, and then it decelerates. The total displacement between the stop signs is equal to the sum of the three displacements, that is, $x_3 - x_0 = (x_3 - x_2) + (x_2 - x_1) + (x_1 - x_0)$.

Known
$x_0 = 0 \quad v_0 = 0$
$t_0 = 0 \quad a_0 = 2.0 \text{ m/s}^2$
$t_1 = 6 \text{ s} \quad t_2 = 8 \text{ s}$
$v_2 = v_1$
$a_2 = -1.5 \text{ m/s}^2$
$v_3 = 0$

Find
x_3

Solve: First, the car accelerates:

$$v_1 = v_0 + a_0(t_1 - t_0) = 0 \text{ m/s} + (2.0 \text{ m/s}^2)(6 \text{ s} - 0 \text{ s}) = 12 \text{ m/s}$$

$$x_1 = x_0 + v_0(t_1 - t_0) + \frac{1}{2}a_0(t_1 - t_0)^2 = 0 \text{ m} + \frac{1}{2}(2.0 \text{ m/s}^2)(6 \text{ s} - 0 \text{ s})^2 = 36 \text{ m}$$

Second, the car moves at v_1:

$$x_2 - x_1 = v_1(t_2 - t_1) + \frac{1}{2}a_1(t_2 - t_1)^2 = (12 \text{ m/s})(8 \text{ s} - 6 \text{ s}) + 0 \text{ m} = 24 \text{ m}$$

Third, the car decelerates:

$$v_3 = v_2 + a_2(t_3 - t_2) \Rightarrow 0 \text{ m/s} = 12 \text{ m/s} + (-1.5 \text{ m/s}^2)(t_3 - t_2) \Rightarrow (t_3 - t_2) = 8 \text{ s}$$

$$x_3 = x_2 + v_2(t_3 - t_2) + \frac{1}{2}a_2(t_3 - t_2)^2 \Rightarrow x_3 - x_2 = (12 \text{ m/s})(8 \text{ s}) + \frac{1}{2}(-1.5 \text{ m/s}^2)(8 \text{ s})^2 = 48 \text{ m}$$

Thus, the total distance between stop signs is

$$x_3 - x_0 = (x_3 - x_2) + (x_2 - x_1) + (x_1 - x_0) = 48 \text{ m} + 24 \text{ m} + 36 \text{ m} = 108 \text{ m}$$

or 110 m to two significant figures.

Assess: A distance of approximately 360 ft in a time of around 16 s with an acceleration/deceleration is reasonable.

P2.73. Prepare: Use the kinematic equations with $(v_x)_i = 0$ m/s in the acceleration phase.

Solve: The man gains speed at a steady rate for the first 1.8 s to reach a top speed of

$$(v_x)_f = (v_x)_i + a_x \Delta t = 0 \text{ m/s} + (6.0 \text{ m/s}^2)(1.8 \text{ s}) = 10.8 \text{ m/s}$$

During this time he will go a distance of

$$\Delta x = \frac{1}{2}a_x(\Delta t)^2 = \frac{1}{2}(6.0 \text{ m/s}^2)(1.8 \text{ s})^2 = 9.72 \text{ m}$$

The man then covers the remaining 100 m − 9.72 m = 90.28 m at constant velocity in a time of

$$\Delta t = \frac{\Delta x}{v_x} = \frac{90.28 \text{ m}}{10.8 \text{ m/s}} = 8.4 \text{ s}$$

The total time for the man is then 1.8 s + 8.4 s = 10.2 s for the 100 m.

We now re-do all the calculations for the horse going 200 m. The horse gains speed at a steady rate for the first 4.8 s to reach a top speed of

$$(v_x)_f = (v_x)_i + a_x\Delta t = 0 \text{ m/s} + (5.0 \text{ m/s}^2)(4.8 \text{ s}) = 24 \text{ m/s}$$

During this time the horse will go a distance of

$$\Delta x = \frac{1}{2}a_x(\Delta t)^2 = \frac{1}{2}(5.0 \text{ m/s}^2)(4.8 \text{ s})^2 = 57.6 \text{ m}$$

The horse then covers the remaining 200 m − 57.6 m = 142.4 m at constant velocity in a time of

$$\Delta t = \frac{\Delta x}{v_x} = \frac{142.2 \text{ m}}{24 \text{ m/s}} = 5.9 \text{ s}$$

The total time for the horse is then 4.8 s + 5.9 s = 10.7 s for the 200 m.

The man wins the race (10.2 s < 10.7 s), but he only went half the distance the horse did.

Assess: We know that 10.2 s is about right for a human sprinter going 100 m. The numbers for the horse also seem reasonable.

P2.75. Prepare: A visual overview of the two cars that includes a pictorial representation, a motion diagram, and a list of values is shown below. We label the motion of the two cars along the x-axis. Constant acceleration kinematic equations are applicable because both cars have constant accelerations. We can easily calculate the times $(t_f)_H$ and $(t_f)_P$ from the given information.

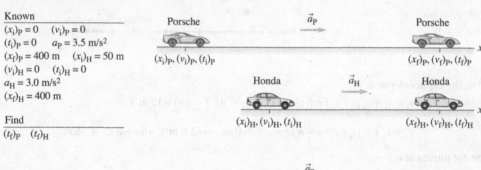

Known	
$(x_i)_P = 0$	$(v_i)_P = 0$
$(t_i)_P = 0$	$a_P = 3.5 \text{ m/s}^2$
$(x_f)_P = 400 \text{ m}$	$(x_i)_H = 50 \text{ m}$
$(v_i)_H = 0$	$(t_i)_H = 0$
$a_H = 3.0 \text{ m/s}^2$	
$(x_f)_H = 400 \text{ m}$	

Find
$(t_f)_P \quad (t_f)_H$

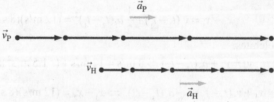

Solve: The Porsche's time to finish the race is determined from the position equation

$$(x_f)_P = (x_i)_P + (v_i)_P((t_f)_P - (t_i)_P) + \frac{1}{2}a_P((t_f)_P - (t_i)_P)^2$$

$$\Rightarrow 400 \text{ m} = 0 \text{ m} + 0 \text{ m} + \frac{1}{2}(3.5 \text{ m/s}^2)((t_f)_P - 0 \text{ s})^2 \Rightarrow (t_f)_P = 15 \text{ s}$$

The Honda's time to finish the race is obtained from Honda's position equation as

$$(x_f)_H = (x_i)_H + (v_i)_H ((t_f)_H - (t_i)_H) + \frac{1}{2} a_H ((t_f)_H - (t_i)_H)^2$$

$$400 \text{ m} = 100 \text{ m} + 0 \text{ m} + \frac{1}{2}(3.0 \text{ m/s}^2)((t_f)_H - 0 \text{ s})^2 \Rightarrow (t_f)_H = 14 \text{ s}$$

The Honda wins by 1.0 s.

Assess: It seems reasonable that the Honda would win given that it only had to go 300 m. If the Honda's head start had only been 50 m rather than 100 m the race would have been a tie.

3

VECTORS AND MOTION IN TWO DIMENSIONS

Q3.3. Reason: Consider two vectors $\vec{A}$ and $\vec{B}$. Their sum can be found using the method of algebraic addition. In Question 3.2 we found that the components of the zero vector are both zero. The components of the resultant of $\vec{A}$ and $\vec{B}$ must then be zero also. So

$$R_x = A_x + B_x = 0$$
$$R_y = A_y + B_y = 0$$

Solving for the components of $\vec{B}$ in terms of $\vec{A}$ gives $B_x = -A_x$ and $B_y = -A_y$. Then the magnitude of $\vec{B}$ is $\sqrt{(B_x)^2 + (B_y)^2} = \sqrt{(-A_x)^2 + (-A_y)^2} = \sqrt{(A_x)^2 + (A_y)^2}$. So then the magnitude of $\vec{B}$ is exactly equal to the magnitude of $\vec{A}$. So, no, two vectors of unequal magnitudes cannot sum to $\vec{0}$.

Assess: For two vectors to add to zero, the vectors must have exactly the same magnitude and point in opposite directions.

Q3.5. Reason: The ones that are constant are v_x, a_x, and a_y. Furthermore, a_x is not only constant, it is zero.

Assess: There are instants when other quantities can be zero, but not throughout the flight. Remember that $a_y = -g$ throughout the flight and that v_x is constant; that is, projectile motion is nothing more than the combination of two simple kinds of motion: constant horizontal velocity and constant vertical acceleration.

Q3.7. Reason: By extending their legs forward, the runners increase their time in the air. As you will learn in chapter 7, the "center of mass" of a projectile follows a parabolic path. By raising their feet so that their feet are closer to their center of mass, the runners increase the time it takes for their feet to hit the ground. By increasing their time of flight, they increase their range. Also, having their feet ahead of them means that their feet will land ahead of where they would have landed otherwise.

Assess: By simply moving their feet, runners can change their time of flight and change the spot where their feet land.

Q3.9. Reason: The greatest range is achieved by using a launch angle of 45°. The greatest "hang time" is achieved by kicking the ball straight up at 90° from the horizontal.

Assess: For a large hang time we want the vertical component of the velocity to be large.

Q3.15. Reason: The magnitude of the acceleration is $\dfrac{v^2}{r}$ which is constant since the speed is constant. So the magnitude of the acceleration doesn't change. But the direction does since it always points toward the center of the circle.

Assess: In circular motion, the acceleration is always toward the center of the circle and is called centripetal acceleration.

Q3.17. Reason: To make a tighter turn with a smaller radius, you need to reduce your speed. If you are traveling at the greatest speed which is safe, then you are accelerating at the highest acceleration which is safe. Now if the radius is reduced, this tends to increase the acceleration above safe values. To bring it back down, your speed should be reduced. If we solve the formula for centripetal acceleration, $a = v^2 / r$, for v, we have: $v = \sqrt{ar}$. So v is proportional to the square root of r. For example, if we take a turn with a radius which is four times smaller, we need to cut our speed in half.

Assess: The maximum safe speed is proportional to the square root of the radius of the turn.

Q3.19. Reason: To generate a vector which points to the left, we could add two vectors which point left, one pointing up and the other down. In C, $\vec{Q}$ and $-\vec{P}$ fit this description so their sum points to the left. The various vector combinations are shown.

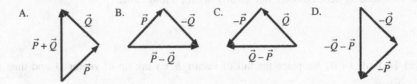

The answer is C.

Assess: If two vectors have equal and opposite components in a certain direction, say the x direction, then when we add the vectors, the equal and opposite components will cancel and leave us with a vector perpendicular to that direction.

Q3.21. Reason: The car is traveling at constant speed, so the only possible cause for accelerations is a change in direction.
(a) At point 1 the car is traveling straight to the right on the diagram, so its velocity is straight to the right. The correct choice is B.
(b) At point 1 the car is traveling at constant speed and not changing direction so its acceleration is zero. The correct choice is E.
(c) The car's velocity at this point on the curve is in the direction of its motion, which is in the direction shown at choice C.
(d) The car is moving on a portion of a circle. The acceleration of an object moving in a circle is always directly toward the center of the circle. The correct choice is D.
(e) The car is moving on a portion of a circle at point 3. The instantaneous velocity vector is directly to the right, which is choice B.
(f) The car is accelerating because it is moving on a portion of a circle. The acceleration is toward the center of the circle, which is in direction A.

Assess: The instantaneous velocity of a particle is always in the direction of its motion at that point in time. For motion in a circle, the direction of the acceleration is always toward the center of the circle.

Q3.25. Reason: The car drops in the vertical direction by a distance of 73 m. Since the car drives off the cliff horizontally, its initial velocity in the vertical direction is 0 m/s. Projectile motion is described by Synthesis 3.1.
(a) To find time given distance we can use

$$y_f = y_i + (v_y)_i \Delta t - \frac{1}{2} g (\Delta t)^2$$

With $y_i = 0$ m, $y_f = -73$ m and $(v_y)_i = 0$ m/s. Substituting these values in and solving for Δt

$$\Delta t = \sqrt{\frac{-2 y_f}{g}} = \sqrt{\frac{-2(-73 \text{ m})}{(9.80 \text{ m/s}^2)}} = 3.9 \text{ s}$$

The correct choice is C.
(b) The horizontal distance traveled by the car is found by multiplying v_x by Δt:

$$v_x \Delta t = (27 \text{ m/s})(3.86 \text{ s}) = 104 \text{ m}$$

Here we have used three figures in Δt since it is used in an intermediate calculation. To two significant figures, the car lands 100 m from the base of the cliff, so the answer is C.

Assess: Projectile motion is made up of two independent motions: uniform motion at constant velocity in the horizontal direction and free-fall motion in the vertical direction. Note that the initial velocity of the car was not relevant to this problem because it was entirely in the horizontal direction.

Q3.29. Reason: The magnitude of the centripetal acceleration is

$$a_c = \frac{v^2}{r} = \frac{(68 \text{ m/s})^2}{95 \text{ m}} = 48.7 \text{ m/s}^2$$

Divide this by 9.8 m/s^2 to find the answer in units of g: $\dfrac{48.7 \text{ m/s}^2}{9.8 \text{ m/s}^2} = 5g$ so the correct choice is E.

Assess: People can withstand $5g$ accelerations, but usually not for a long time.

Problems

P3.1. Prepare: (a) To find $\vec{A} + \vec{B}$, we place the tail of vector $\vec{B}$ on the tip of vector $\vec{A}$ and then connect vector $\vec{A}$'s tail with vector $\vec{B}$'s tip.

(b) To find $\vec{A} - \vec{B}$, we note that $\vec{A} - \vec{B} = \vec{A} + (-\vec{B})$. We place the tail of vector $-\vec{B}$ on the tip of vector $\vec{A}$ and then connect vector $\vec{A}$'s tail with the tip of vector $-\vec{B}$.

Solve:

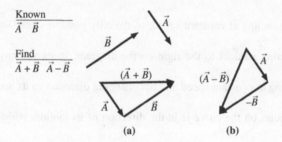

P3.3. Prepare: We can find the positions and velocity and acceleration vectors using a motion diagram.

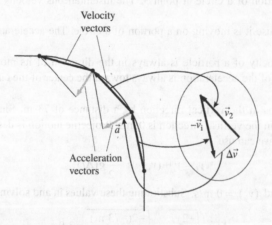

Solve: The figure gives several points along the car's path. The velocity vectors are obtained by connecting successive dots. The acceleration vectors are obtained by subtracting successive velocity vectors. The acceleration vectors point toward the center of the diagram.

Assess: Notice that the acceleration points toward the center of the turn. As you will learn in chapter 4, whenever your car accelerates, you feel like you are being pushed the opposite way. This is why you feel like you are being pushed away from the center of a turn.

P3.7. Prepare: The figure below shows the components v_x and v_y, and the angle θ. We will use Tactics Box 3.3 to find the sign attached to the components of a vector.

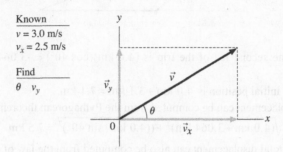

Solve: **(a)** Since $v_x = v \cos \theta$, we have 2.5 m/s $= (3.0$ m/s$) \cos \theta \Rightarrow \theta = \cos^{-1}(2.5$ m/s/30 m/s$) = 33.6°$ or 34 degrees to two significant figures.
(b) The vertical component is $v_y = v \sin \theta = (3.0$ m/s$) \sin 33.6° = 1.7$ m/s.

P3.11. Prepare: We will follow rules given in Tactics Box 3.3.

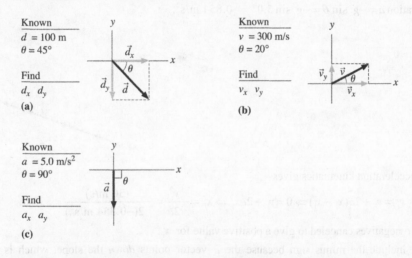

Solve: **(a)** Vector $\vec{d}$ points to the right and down, so the components d_x and d_y are positive and negative, respectively:

$$d_x = d \cos \theta = (100 \text{ m}) \cos 45° = 71 \text{ m} \quad d_y = -d \sin \theta = -(100 \text{ m}) \sin 45° = -71 \text{ m}$$

(b) Vector $\vec{v}$ points to the right and up, so the components v_x and v_y are both positive:

$$v_x = v \cos \theta = (300 \text{ m/s}) \cos 20° = 280 \text{ m/s} \quad v_y = v \sin \theta = (300 \text{ m/s}) \sin 20° = 100 \text{ m/s}$$

(c) Vector $\vec{a}$ has the following components:

$$a_x = -a \cos \theta = -(5.0 \text{ m/s}^2) \cos 90° = 0.0 \text{ m/s}^2 \quad a_y = -a \sin \theta = -(5.0 \text{ m/s}^2) \sin 90° = -5.0 \text{ m/s}^2$$

Assess: The components have the same units as the vectors. Note the minus signs we have manually inserted according to Tactics Box 3.3.

P3.15. Prepare: Draw a diagram of the situation.

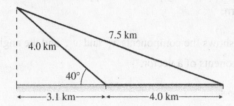

Solve: The x-component of the second leg of the trip is $(4.0 \text{ km})(\cos 40°) = -3.064$ km which is about 3.1 km more west.

The total distance west of their initial position is $4.0 \text{ km} + 3.1 \text{ km} = 7.1$ km.

The magnitude of the total displacement can be computed from the Pythagorean theorem:
$$\sqrt{(4.0 \text{ km} + 3.064 \text{ km})^2 + ((4.0 \text{ km}) \sin 40°)^2} = 7.5 \text{ km}$$

Assess: The magnitude of the total displacement can also be computed from the law of cosines:
$$d^2 = (4.0 \text{ km})^2 + (4.0 \text{ km})^2 - 2(4.0 \text{ km})(4.0 \text{ km}) \cos 140° \Rightarrow d = 7.5 \text{ km}$$

P3.17. Prepare: A visual overview of the car's motion that includes a pictorial representation, a motion diagram, and a list of values is shown below. We have labeled the x-axis along the incline. Note that the problem "ends" at a turning point, where the car has an instantaneous speed of 0 m/s before rolling back down. The rolling back motion is *not* part of this problem. If we assume the car rolls without friction, then we have motion on a frictionless inclined plane with acceleration $a = -g \sin \theta = -g \sin 5.0° = -0.854 \text{ m/s}^2$.

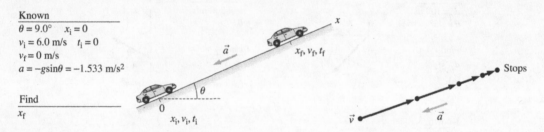

Solve: Constant acceleration kinematics gives
$$v_f^2 = v_i^2 + 2a(x_f - x_i) \Rightarrow 0 = v_i^2 + 2ax_f \Rightarrow x_f = -\frac{v_i^2}{2a} = -\frac{(30 \text{ m/s})^2}{2(-0.854 \text{ m/s}^2)} = 530 \text{ m}$$

Notice how the two negatives canceled to give a positive value for x_f.

Assess: We must include the minus sign because the $\vec{a}$ vector points *down* the slope, which is in the negative x-direction.

P3.21. Prepare: For everyday speeds we can use Equation 3.21 to find relative velocities. We will use a subscript A for Anita and a 1 and a 2 for the respective balls; we also use a subscript G for the ground. We will consider all motion in this problem to be along the x-axis (ignore the vertical motion including the fact that the balls also fall under the influence of gravity) and so we drop the x subscript.

It is also worth noting that interchanging the order of the subscripts merely introduces a negative sign. For example, $v_{AG} = 5$ m/s, so $v_{GA} = -5$ m/s.

"According to Anita" means "relative to Anita."

Solve: For ball 1:
$$v_{1A} = v_{1G} + v_{GA} = 10 \text{ m/s} + (-5 \text{ m/s}) = 5 \text{ m/s}$$

For ball 2:
$$v_{2A} = v_{2G} + v_{GA} = -10 \text{ m/s} + (-5 \text{ m/s}) = -15 \text{ m/s}$$

The speed is the magnitude of the velocity, so the speed of ball 2 is 15 m/s.

Assess: You can see that at low speeds velocities simply add or subtract, as the case may be. Mentally put yourself in Anita's place, and you will confirm that she sees ball 1 catching up to her at only 5 m/s while she sees ball 2 speed past her at 15 m/s.

P3.23. Prepare: We can use the technique of "canceling" subscripts to find relative velocities.

Solve: Anita's friends are standing on the ground, so we can calculate the velocities they threw the balls with by calculating the velocities of the balls relative to the ground. The velocity of ball 1 relative to Anita is $(v_x)_{1A} = +10$ m/s. The velocity of ball 2 relative to Anita is $(v_x)_{2A} = -10$ m/s. Anita's velocity relative to the ground is $(v_x)_{Ag} = +5$ m/s. Then the velocity of ball 1 relative to the ground is

$$(v_x)_{1g} = (v_x)_{1A} + (v_x)_{Ag} = +10 \text{ m/s} + 5 \text{ m/s} = +15 \text{ m/s}$$

The velocity of ball 2 relative to the ground is

$$(v_x)_{2g} = (v_x)_{2A} + (v_x)_{Ag} = -10 \text{ m/s} + 5 \text{ m/s} = -5 \text{ m/s}$$

The speed is the magnitude of the velocity, so the speed of ball 2 is 5 m/s.

Assess: The results make sense. The ball to the left of Anita must be traveling faster than Anita, and the ball to the right must be traveling slower than Anita.

P3.25. Prepare: Assume motion along the x-direction. The velocity of the boat relative to the ground is $(v_x)_{bg}$; the velocity of the boat relative to the water is $(v_x)_{bw}$; and the velocity of the water relative to the ground is $(v_x)_{wg}$. We will use the technique of Equation 3.21: $(\vec{v}_x)_{bg} = (\vec{v}_x)_{bw} + (\vec{v}_x)_{wg}$.

Solve: For travel down the river,

$$(v_x)_{bg} = (v_x)_{bw} + (v_x)_{wg} = \frac{30 \text{ km}}{3.0 \text{ h}} = 10.0 \text{ km/h}$$

For travel up the river,

$$(v_x)_{bg} = -(v_x)_{bw} + (v_x)_{wg} = -\left(\frac{30 \text{ km}}{5.0 \text{ h}}\right) = -6.0 \text{ km/h}$$

Adding these two equations yields $(v_x)_{wg} = 2.0$ km/h. That is, the velocity of the flowing river relative to the earth is 2.0 km/h.

Assess: Note that the speed of the boat relative to the water downstream and upstream are the same.

P3.27. Prepare: We will assume the ball is in free fall (*i.e.*, we neglect air resistance). The trajectory of a projectile is a parabola because it is a combination of constant horizontal velocity ($a_x = 0.0$ m/s^2) combined with constant vertical acceleration ($a_y = -g$). In this case we see only half of the parabola.

The initial speed given is all in the horizontal direction, that is, $(v_x)_i = 5.0$ m/s and $(v_y)_i = 0.0$ m/s.

Solve:

(a) 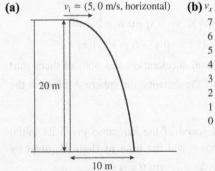   **(b)** 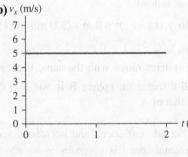   **(c)**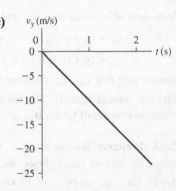

(d) This is a two-step problem. We first use the vertical direction to determine the time it takes, then plug that result into the equation for the horizontal direction.

$$\Delta y = \frac{1}{2} a_y (\Delta t)^2$$

$$\Delta t = \sqrt{\frac{2\Delta y}{a_y}} = \sqrt{\frac{2(-20 \text{ m})}{-9.8 \text{ m/s}^2}} = 2.0 \text{ s}$$

We use the 2.0 s in the equation for the horizontal motion.

$$\Delta x = v_x \Delta t = (5.0 \text{ m/s})(2.0 \text{ s}) = 10 \text{ m}$$

Assess: The answers seem reasonable, and we would get the same answers to two significant figures in a quick mental calculation using $g \approx 10 \text{ m/s}^2$. In fact, I did this before computing the algebra so I would know how to scale the graphs.

P3.31. Prepare: We will apply the constant-acceleration kinematic equations to the horizontal and vertical motions as described by Synthesis 3.1.

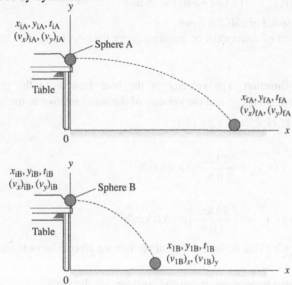

Known
$x_{iA} = x_{iB} = t_{iA} = t_{iB} = 0$
$y_{iA} = y_{iB} = 1.0 \text{ m}$
$y_{fA} = y_{fB} = 0$
$m_A = 1.0 \text{ kg} \quad m_B = 0.4 \text{ kg}$
$(v_x)_{iA} = 5.0 \text{ m/s}$
$(v_x)_{iB} = 2.5 \text{ m/s}$
$(v_y)_{iA} = (v_y)_{iB} = 0$
$a_y = -g$

Find
$t_{fA} \quad t_{fB}$
$x_{fA} \quad x_{fB}$

Solve: Using $y_{fA} = y_{iA} + (v_y)_{iA}(t_{fA} - t_{iA}) + \frac{1}{2}(a_y)_A (t_{fA} - t_{iA})^2$, we get

$$-1.0 \text{ m} = 0 \text{ m} + 0 \text{ m} + \frac{1}{2}(-9.8 \text{ m/s}^2)(t_{fA} - 0 \text{ s})^2 \Rightarrow t_{fA} = 0.452 \text{ s} = t_{fB}$$

This should be reported as 0.45 s to two significant figures. Because $y_{fA} = y_{fB}$, both take the same time to reach the floor.

We are now able to calculate x_{fA} and x_{fB} as follows:

$$x_{fA} = x_{iA} + (v_x)_{iA}(t_{fA} - t_{iA}) + \frac{1}{2}(a_x)_A (t_{fA} - t_{iA})^2 = 0 \text{ m} + (5.0 \text{ m/s})(0.452 \text{ s} - 0 \text{ s}) + 0 \text{ m} = 2.3 \text{ m}$$

$$x_{fB} = x_{iB} + (v_x)_{iB}(t_{fB} - t_{iB}) + \frac{1}{2}(a_x)_B (t_{fB} - t_{iB})^2 = 0 \text{ m} + (2.5 \text{ m/s})(0.452 \text{ s} - 0 \text{ s}) + 0 \text{ m} = 1.1 \text{ m}$$

Assess: Note that $t_{fB} = t_{fA}$ since both the spheres move with the same vertical acceleration and both of them start with zero vertical velocity. The horizontal distance for sphere B is one-half the distance for sphere A because the horizontal velocity of sphere B is one-half that of A.

P3.33. Prepare: We are asked to find the take-off speed and horizontal speed of the kangaroo given its initial angle, 20°, and its range. Since the horizontal speed is given by $v_x = v \cos \theta$ and the time of flight is given by $\Delta t = 2v \sin \theta / g$, the range of the kangaroo is given by the product of these: $\Delta x = 2v \sin \theta \cos \theta / g$.

Solve: (a) We can solve the above formula for v and then plug in the range and angle to find the take-off speed:

$$v = g\Delta x / (2 \sin \theta \cos \theta) = (9.8 \text{ m/s}^2)(10 \text{ m})/(2 \sin 20° \cos 20°) = 12.3 \text{ m/s}$$

Its take-off speed is 12 m/s, to two significant figures.

(b) Its horizontal speed is given by $v_x = v\cos\theta = (12.3 \text{ m/s})\cos 20° = 11.6 \text{ m/s}$ or 12 m/s to two significant figures.

Assess: The reason the horizontal speed and take-off speed appear the same is that $20°$ is a small angle and the cosine of a small angle is approximately equal to 1.

P3.39. Prepare: We can use the formula for centripetal acceleration, Equation 3.30. We need the radius of the track which is half the diameter: $R = D/2 = (8.0 \text{ m})/2 = 4.0 \text{ m}$.

Solve:

$$a = \frac{v^2}{r} = \frac{(5.0 \text{ m/s})^2}{4.0 \text{ m}} = 6.3 \text{ m/s}^2$$

Assess: This is a fairly large acceleration for a car.

P3.41. Prepare: Examine the formula carefully.

$$a = \frac{v^2}{r}$$

Before plugging in numbers, notice that if the speed is held constant (as in part **(a)**), then a and r are inversely proportional to each other: doubling one halves the other. And if r is held constant (as in part **(b)**), then there is a square relationship between a and v: doubling v quadruples a.

Solve: It is convenient to use ratios to solve this problem, because we never have to know any specific values for r or v. We'll use unprimed variables for the original case ($a = 8.0 \text{ m/s}^2$), and primed variables for the new cases.

(a) With the speed held constant, $v' = v$ but $r' = 2r$.

$$\frac{a'}{a} = \frac{\frac{v'^2}{r'}}{\frac{v^2}{r}} = \frac{\frac{v^2}{2r}}{\frac{v^2}{r}} = \frac{1}{2}$$

So $a' = \frac{1}{2}a = \frac{1}{2}(8.0 \text{ m/s}^2) = 4.0 \text{ m/s}^2$.

(b) With the radius held constant, $r'' = r$ but $v'' = 2v$.

$$\frac{a''}{a} = \frac{\frac{v''^2}{r''}}{\frac{v^2}{r}} = \frac{\frac{(2v)^2}{r}}{\frac{v^2}{r}} = 2^2 = 4$$

So $a'' = 4a = 4(8.0 \text{ m/s}^2) = 32 \text{ m/s}^2$.

Assess: Please familiarize yourself with this ratio technique and look for opportunities to use it. The advantage is not needing to know any specific values of r or v—not only not having to know them, but realizing that the result is independent of them.

The daily life lesson is that driving around a curve with a larger radius produces gentler acceleration. Going around a given curve faster, however, requires a much larger acceleration (produced by the friction between the tires and the road), and the relationship is squared. If your tires are bald or the road slippery there won't be enough friction to keep you on the road if you go too fast. And remember that it is a squared relationship, so going around a curve twice as fast requires four times as much friction.

P3.43. Prepare: The magnitude of centripetal acceleration is given in Equation 3.30.

Solve: The centripetal acceleration is given as 1.5 times the acceleration of gravity, so

$$a = (1.5)(9.80 \text{ m/s}^2) = 15 \text{ m/s}^2$$

Using Equation 3.30, the radius of the turn is given by

$$r = \frac{v^2}{a} = \frac{(20 \text{ m/s})^2}{15 \text{ m/s}^2} = 27 \text{ m}$$

Assess: This seems reasonable.

P3.45. Prepare: The vectors $\vec{A}$, $\vec{B}$, and $\vec{D} = \vec{A} - \vec{B}$ are shown. Because $\vec{A} = \vec{A}_x + \vec{A}_y$ and $\vec{B} = \vec{B}_x + \vec{B}_y$, so the components of the resultant vector are $D_x = A_x - B_x$ and $D_y = A_y - B_y$.

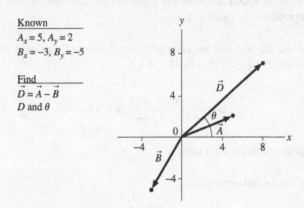

Known
$A_x = 5, A_y = 2$
$B_x = -3, B_y = -5$

Find
$\vec{D} = \vec{A} - \vec{B}$
D and θ

Solve: (a) With $A_x = 5$, $A_y = 2$, $B_x = -3$, and $B_y = -5$, we have $D_x = 8$ and $D_y = 7$.

(b) Vectors $\vec{A}$, $\vec{B}$ and $\vec{D}$ are shown in the above figure.

(c) Since $D_x = 8$ and $D_y = 7$, the magnitude and direction of $\vec{D}$ are

$$D = \sqrt{(8)^2 + (7)^2} = 11 \qquad \theta = \tan^{-1}\left(\frac{D_y}{D_x}\right) = \tan^{-1}\left(\frac{7}{8}\right) = 41°$$

Assess: Since $|D_y| < |D_x|$, the angle θ is less than $45°$, as it should be.

P3.51. Prepare: We draw a picture of the plane's path. The distance traveled by the plane is given by speed multiplied by time and forms the hypotenuse. The shortest distance to the equator is 100 km and this is the length of the side adjacent to the $30°$ angle. We can relate these two sides of a triangle using trigonometry.

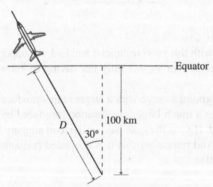

Solve: If D is the distance traveled by the airplane, then we can write $\cos 30° = (100 \text{ km})/D$ and solve for the distance traveled:

$$D = (100 \text{ km})/\cos 30° = 115 \text{ km}$$

The time of flight is equal to the distance traveled divided by speed:

$$\Delta t = (115 \text{ km})/(150 \text{ km/h}) = 0.767 \text{ h}$$

Finally, we convert this time to minutes:

$$0.767 \text{ h} = 0.767 \text{ h}\left(\frac{60 \text{ min}}{1 \text{ h}}\right) = 46 \text{ min}$$

It takes the pilot 46 min to reach the equator.

Assess: The pilot's time to reach the equator is greater than it would be if he were flying directly toward the equator. If the flight were direct, the distance would be 100 km instead of 115 km and the time of flight would be 40 min instead of 46 min.

P3.53. Prepare: The skier's motion on the horizontal, frictionless snow is not of any interest to us. The skier's speed increases down the incline due to acceleration parallel to the incline, which is equal to $g \sin 10°$. A visual overview of the skier's motion that includes a pictorial representation, a motion representation, and a list of values is shown.

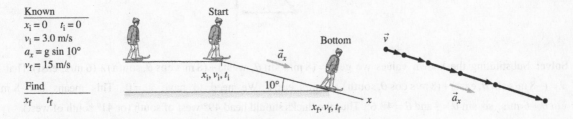

Solve: Using the following constant-acceleration kinematic equations,

$$v_f^2 = v_i^2 + 2a_x(x_f - x_i)$$

$$\Rightarrow (15 \text{ m/s})^2 = (3.0 \text{ m/s})^2 + 2(9.8 \text{ m/s}^2)\sin 10°(x_f - 0 \text{ m}) \Rightarrow x_f = 63 \text{ m}$$

$$v_f = v_i + a_x(t_f - t_i)$$

$$\Rightarrow (15 \text{ m/s}) = (3.0 \text{ m/s}) + (9.8 \text{ m/s}^2)(\sin 10°)t_f \Rightarrow t_f = 7.1 \text{ s}$$

Assess: A time of 7.1 s to cover 63 m is a reasonable value.

P3.55. Prepare: Assume motion along the x-direction. Let $x_f - x_i$ be the displacement from your gate to the baggage claim. We will use the technique of Equation 3.21: $(\vec{v}_x)_{yg} = (\vec{v}_x)_{ym} + (\vec{v}_x)_{mg}$.

Solve: In the first case, when the moving sidewalk is broken, we can find your velocity

$$v_Y = \frac{(x_f - x_i)}{50 \text{ s}}$$

In the second case, when you stand on the moving sidewalk, the velocity of the sidewalk relative to the ground is

$$v_{sg} = \frac{x_f - x_i}{75 \text{ s}}$$

In the third case, when you walk while riding, we can use the equation

$$(v)_{yg} = (v)_{ys} + (v)_{sg}$$

That is, your velocity relative to the ground, when you are walking on the moving sidewalk, is equal to your velocity relative to the moving sidewalk (which is v_Y) plus the sidewalk's velocity relative to the ground. Thus,

$$\frac{x_1 - x_0}{\Delta t} = \frac{x_1 - x_0}{50 \text{ s}} + \frac{x_1 - x_0}{75 \text{ s}} \Rightarrow \Delta t = 30 \text{ s}$$

Assess: A time smaller than 50 s was expected.

P3.57. Prepare: A visual overview of the ducks' motion is shown below. The resulting velocity is given by $\vec{v} = \vec{v}_{fly} + \vec{v}_{wind}$, where $\vec{v}_{wind} = 6$ m/s, east) and $\vec{v}_{fly} = (v_{fly} \sin \theta, \text{west}) + (v_{fly} \cos \theta, \text{south})$.

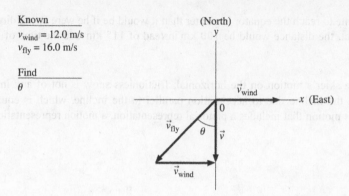

Known
$v_{wind} = 12.0$ m/s
$v_{fly} = 16.0$ m/s

Find
θ

Solve: Substituting the known values we get $\vec{v} = (8$ m/s $\sin\theta$, west$) + (8$ m/s $\cos\theta$, south$) + (6$ m/s, east$)$. That is, $\vec{v} = (-8$ m/s $\sin\theta$, east$) + (8$ m/s $\cos\theta$, south$) + (6$ m/s, east$)$. We need to have $v_x = 0$. This means $0 = -8$ m/s $\sin\theta + 6$ m/s, so $\sin\theta = \frac{6}{8}$ and $\theta = 48.6°$. Thus the ducks should head 49° west of south (or 41° south of west).

P3.63. Prepare: This problem is somewhat similar to Problem 3.27 with all of the initial velocity in the horizontal direction. We will use the vertical equation for constant acceleration to determine the time of flight and then see how far Captain Brady can go in that time.

Of interest is the fact that we will do this two-step problem completely with variables in part **(a)** and only plug in numbers in part **(b)**.

We *could* do part **(b)** in feet (using $g = 32$ ft/s^2), but to compare with the world record 100 m dash, let's convert to meters. $L = 22$ ft $= 6.71$ m and $h = 20$ ft $= 6.10$ m.

Solve: (a) Given that $(v_y) = 0.0$ ft/s we can use the kinematic equations.

$$(y_f - y_i) = \frac{1}{2}a_y(\Delta t)^2$$

With up as the positive direction, $(y_f - y_i)$ is negative and $a_y = -g$; those signs cancel leaving

$$h = \frac{1}{2}g(\Delta t)^2$$

Solve for Δt.

$$\Delta t = \sqrt{\frac{2h}{g}}$$

Now use that expression for Δt in the equation for constant horizontal velocity.

$$L = \Delta x = v_x \Delta t = v_x \sqrt{\frac{2h}{g}}$$

Finally solve for $v = v_x$ in terms of L and h.

$$v = \frac{L}{\sqrt{\frac{2h}{g}}} = L\sqrt{\frac{g}{2h}}$$

Now plug in the numbers we are given for L and h.

$$v = L\sqrt{\frac{g}{2h}} = (6.71 \text{ m})\sqrt{\frac{9.8 \text{ m/s}^2}{2(6.10 \text{ m})}} = 6.0 \text{ m/s}$$

(b) Compare this result ($v = 6.0$ m/s) with the world-class sprinter ($v = 10$ m/s); a fit person could make this leap.

Assess: The results are reasonable, and not obviously wrong. 6.0 m/s $\approx$ 13 mph, and that would be a fast run, but certainly possible.

By solving the problem first algebraically before plugging in any numbers, we are able to substitute other numbers as well, if we desire, without re-solving the whole problem.

P3.65. Prepare: We will use the initial information (that the marble goes 6.0 m straight up) to find the speed the marble leaves the gun. We also need to know how long it takes something to fall 1.5 m from rest in free fall so we can then use that in the horizontal equation.

Assume that there is no air resistance ($a_y = -g$) and that the marble leaves the gun with the same speed (muzzle speed) each time it is fired.

Solve: To determine the muzzle speed in the straight-up case, use Equation 2.13.

$$(v_y)_f^2 = (v_y)_i^2 + 2a_y \Delta y$$

where at the top of the trajectory $(v_y)_f = 0.0$ m/s and $\Delta y = 6.0$ m.

$$(v_y)_i^2 = 2g\Delta y \Rightarrow (v_y)_i = \sqrt{2g\Delta y} = 10.8 \text{ m/s}$$

We also rearrange Equation 2.12 to find the time for an object to fall 1.5 m from rest: $y_f - y_i = -15$ m now instead of the 6.0 m used previously.

$$\Delta y = \frac{1}{2} a_y (\Delta t)^2$$

$$\Delta t = \sqrt{\frac{2\Delta y}{-g}} = \sqrt{\frac{2(-1.5 \text{ m})}{-9.8 \text{ m/s}^2}} = 0.553 \text{ s}$$

At last we combine this information into the equation for constant horizontal velocity.

$$\Delta x = v_x \Delta t = (10.8 \text{ m/s})(0.553 \text{ s}) = 6.0 \text{ m}$$

Assess: Is it a coincidence that the marble has a horizontal range of 6.0 m when it can reach a height of 6.0 m when fired straight up, or will those numbers always be the same? Well, the 6.0 m horizontal range depends on the height (1.5 m) from which you fire it, so if that were different the range would be different. This leads us to conclude that it *is* a coincidence. You can go back, though, and do the problem algebraically (with no numbers) and find that g cancels and that the horizontal range is 2 times the square root of the product of the vertical height it can reach and the height from which you fire it horizontally.

P3.67. Prepare: We will apply the constant-acceleration kinematic equations to the horizontal and vertical motions of the shot as described by Synthesis 3.1. A visual overview is shown.

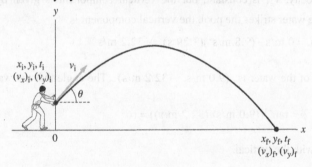

Known	
$x_i = t_i = 0$	$y_i = 1.8$ m
$v_i = 12$ m/s	$\theta = 40°$
$a_y = -g$	$(v_x)_i = (v_x)_f$

Find
x_f

Solve: (a) Using $y_f = y_i + (v_y)_i(t_f - t_i) + \frac{1}{2}a_y(t_f - t_i)^2$,

$$0 \text{ m} = 1.8 \text{ m} + v_i \sin 40°(t_f - 0 \text{ s}) + \frac{1}{2}(-9.8 \text{ m/s}^2)(t_f - 0 \text{ s})^2$$

$$= 1.8 \text{ m} + (7.713 \text{ m/s})t_f - (4.9 \text{ m/s}^2)t_f^2 \Rightarrow t_f = -0.206 \text{ s and } 1.780 \text{ s}$$

The negative value of t_f is unphysical for the current situation. Using $t_f = 1.780$ s and $x_f = x_i + (v_x)_i(t_f - t_i)$, we get

$$x_f = 0 + (v_i \cos 40° \text{ m/s})(1.780 \text{ s} - 0 \text{ s}) = (12 \text{ m/s}) \cos 40°(1.78 \text{ s}) = 16.36 \text{ m} = 16.4 \text{ m}$$

(b) We can repeat the calculation for each angle. A general result for the flight time at angle q is

$$t_f = (12 \sin\theta + \sqrt{144\sin^2\theta + 35.28})/9.8 \text{ s}$$

and the distance traveled is $x_f = 12 \cos\theta \times t_f$. We can put the results in a table.

θ	t_f	x_f
40.0°	1.780 s	16.36 m
42.5°	1.853 s	16.39 m
45.0°	1.923 s	16.31 m
47.5°	1.990 s	16.13 m

Maximum distance is achieved at $\theta \approx 42.5°$.

Assess: The well-known "fact" that maximum distance is achieved at 45° is true only when the projectile is launched and lands at the *same* height. That isn't true here. The extra 0.03 m = 3 cm obtained by increasing the angle from 40.0° to 42.5° could easily mean the difference between first and second place in a world-class meet.

P3.71. Prepare: We can use the equation for vertical motion at constant acceleration to find the time of fall and then use the time to find the final velocity.

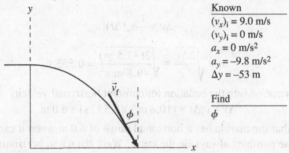

Known
$(v_x)_i = 9.0$ m/s
$(v_y)_i = 0$ m/s
$a_x = 0$ m/s^2
$a_y = -9.8$ m/s^2
$\Delta y = -53$ m

Find
ϕ

Solve: Since the water is launched horizontally, its time of flight and vertical displacement are related by the equation: $\Delta y = -\frac{1}{2}g\Delta t^2$. Solving for the time, we have

$$\Delta t = \sqrt{2|\Delta y|/g} = \sqrt{2(53 \text{ m})/9.8 \text{ m/s}^2} = 3.29 \text{ s}$$

The horizontal component of the velocity, v_x, is constant, but the vertical component is given by the equation: $(v_y)_f = (v_y)_i + a_y\Delta t$. At the moment the water strikes the pool, the vertical component is

$$(v_y)_f = 0 \text{ m/s} - (9.8 \text{ m/s}^2)(3.29 \text{ s}) = -32.2 \text{ m/s}$$

At the moment of impact the velocity of the water is: $(9.0 \text{ m/s}, \ -32.2 \text{ m/s})$. The angle that the water makes with the vertical is given by

$$\phi = \tan^{-1}((9.0 \text{ m/s})/(32.2 \text{ m/s})) = 16°$$

The water is falling at an angle of 16° with the vertical.

Assess: Even though the water is launched at a fairly high speed (9.0 m/s is about 20 mi/hr), it is close to the vertical when it lands because it spends such a long time in the air during which time the absolute value of v_x increases steadily.

P3.75. Prepare: We need to convert the radius of the Mini Cooper's turn to meters and convert the final speed of the Mustang to meters per second. The radius of the Mini Cooper's turn is the following:

$$17 \text{ ft} = 17 \text{ ft}\left(\frac{1 \text{ m}}{3.28 \text{ ft}}\right) = 5.18 \text{ m}$$

And the final speed of the Mustang is as follows:

$$60 \frac{\text{mi}}{\text{h}} = 60 \frac{\text{mi}}{\text{h}}\left(\frac{1609 \text{ m}}{1 \text{ mi}}\right)\left(\frac{1 \text{ h}}{3600 \text{ s}}\right) = 26.8 \text{ m/s}$$

The acceleration of the Mustang is given by $a = \Delta v / \Delta t$

$$a = (26.8 \text{ m/s})/(5.6 \text{ s}) = 4.79 \text{ m/s}^2$$

Solve: To match the Mustang's acceleration, the Mini Cooper must have a centripetal acceleration of 4.79 m/s^2. Given the formula for centripetal acceleration, $a = v^2/r$, we can solve for the necessary radius as follows:

$$v = \sqrt{ar} = \sqrt{(4.79 \text{ m/s}^2)(5.18 \text{ m})} = 4.98 \text{ m/s} = 4.98 \frac{\text{m}}{\text{s}} \left(\frac{1 \text{ mi}}{1609 \text{ m}} \right) \left(\frac{3600 \text{ s}}{1 \text{ h}} \right) = 11 \text{ mph}$$

The Mini Cooper must travel at $5.0 \text{ m/s} = 11$ mph to have the same acceleration as the Mustang.

Assess: Even at a fairly low speed, 11 mph, the acceleration is high. This is because the radius of the turn is so small—17 ft.

P3.77. Prepare: We will use Equation 3.30 to relate the acceleration to the speed. But first we need to convert the speed of the car to m/s.

$$40 \frac{\text{mi}}{\text{hr}} = 40 \frac{\text{mi}}{\text{hr}} \left(\frac{1609 \text{m}}{1 \text{ mi}} \right) \left(\frac{1 \text{ hr}}{3600 \text{ s}} \right) = 17.9 \text{ m/s}$$

Solve: **(a)** Your acceleration is given from the equation $a = v^2/r$

$$\frac{(17.9 \text{ m/s})^2}{110 \text{ m}} = 2.91 \text{ m/s}^2$$

which converts as follows:

$$2.91 \text{ m/s}^2 = (2.91 \text{ m/s}^2) \left(\frac{1g}{9.8 \text{ m/s}^2} \right) = 0.30g$$

The acceleration is 2.9 m/s^2 or $0.30g$.

(b) The formula for centripetal acceleration, $a = v^2/r$ can be solved for v as follows: $v = \sqrt{ar}$. In this form we see that if the acceleration is doubled, then the velocity is multiplied by $\sqrt{2}$. So we multiply the 40 mph speed limit by $\sqrt{2}$: $(40 \text{ mph})\sqrt{2} = 57$ mph. At 57 mph the acceleration would be twice the acceleration at 40 mph.

Assess: As noted in the solution to Problem 42, a small change in velocity can produce a large change in centripetal acceleration. Here, with an increase in speed of less than 50%, the acceleration doubles and the friction needed for the turn also doubles.

FORCES AND NEWTON'S LAWS OF MOTION

Q4.3. Reason: No. If you know all of the forces than you know the direction of the acceleration, not the direction of the motion (velocity). For example, a car moving forward could have on it a net force forward if speeding up or backward if slowing down or no net force at all if moving at constant speed.
Assess: Consider carefully what Newton's *second* law says, and what it doesn't say. The net force must *always* be in the direction of the acceleration. This is also the direction of the *change* in velocity, although not necessarily in the direction of the velocity itself.

Q4.5. Reason: The picture on the left is more effective at tightening the head because of the greater inertia of the head. Once moving, the head will "want" to continue moving (Newton's first law) after the handle hits the table, thus tightening the head, more so than in the second picture where the light handle has less inertia moving down than the head.
Assess: Newton's first law, the law of inertia, says the greater the mass of an object the more it will tend to continue with its previous velocity. One can assess this by trying it with a real hammer with a loose head.

Q4.9. Reason: Since there is no source of gravity, you will not be able to feel the weight of the objects. However, Newton's second law is true even in an environment without gravity. Assuming you can exert a reproducible force in throwing both objects, you could throw each and note the acceleration each obtains.
Assess: Mass is independent of the force of gravity and exists even in environments with no sources of gravity.

Q4.11. Reason: If the ejected air is directed forward then thrust force is backward (Newton's 3rd law). This might be desirable to slow the plane down.
Assess: If the ejected air is directed downward, the thrust force is up. Jets can take off vertically without needing a runway this way.

Q4.13. Reason: The force of Josh on Taylor and the force of Taylor on Josh are members of an action/reaction pair, so that the magnitudes of these two forces are the same. However, since Josh is more massive (bigger) than Taylor, his resulting acceleration during the push will be less and hence his final velocity after the push will be less.
Assess: This problem required a correct conceptual understanding of Newton's second and third laws. The third law allows us to conclude that each skater experiences the same force and the second law allows us to understand that the acceleration is inversely proportional to the mass being accelerated.

Q4.17. Reason: The two forces in question are both exerted on the same object, the filing cabinet. An action/reaction pair is exerted on two different objects. As Alyssa pushes to the right on the filing cabinet, the filing cabinet pushes on Alyssa an equal amount to the left. These two forces constitute an action/reaction pair – note that one force acts on Alyssa and the other force acts on the filing cabinet. The two forces of the action/reaction pair act on different objects. As the frictional force of the floor pushes on the filing cabinet to the left, the filing cabinet pushes on the floor an equal amount to the right. These two forces also constitute an action/reaction pair.
The answer to the question is no, the two forces on the cabinet do not form an action/reaction pair.
Assess: It is important to remember that the forces of an action/reaction pair act on different objects.

Q4.19. Reason: The way the tire is twisted indicated the force of the road on the tire is forward. Since this force is likely greater than the backward air resistance force, the net force is also forward; therefore the car is accelerating in the forward direction. This means it must be speeding up.
Assess: If the car were slowing down the net force would point backward (and the indicative wrinkles in the tire would go the other way).

Q4.23. Reason: Drag points opposite to the direction of motion. As the ball is going up, the drag force acts downward. As the ball comes down, the drag force acts upward. The correct choice is D.
Assess: Drag always acts opposite to the direction of motion of an object.

Q4.25. Reason: The direction of the kinetic friction force will be opposite the motion, so the friction points down while the box goes up, and the friction points up while the box slides down.
The answer is D.
Assess: Drawing a free-body diagram (with tilted axes) and applying Newton's second law will support this conclusion.

Q4.27. Reason: To remain stationary there needs to be a zero net force on the scallop. The downward gravitational force is not quite balanced by the upward buoyant force so the thrust force must also be up. For the thrust force on the scallop to be up, it must eject water in the downward direction.
The answer is C.
Assess: Drawing a free-body diagram (with tilted axes) and applying Newton's second law will support this conclusion.

Problems

P4.5. Prepare: Draw the vector sum $\vec{F}_1 + \vec{F}_2$ of the two forces $\vec{F}_1$ and $\vec{F}_2$. Then look for a vector that will "balance" the force vector $\vec{F}_1 + \vec{F}_2$.

Solve: The object will be in equilibrium if $\vec{F}_3$ has the same magnitude as $\vec{F}_1 + \vec{F}_2$ but is in the opposite direction so that the sum of all three forces is zero.

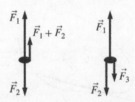

Assess: Adding the new force vector $\vec{F}_3$ with length and direction as shown will cause the object to be at rest.

P4.7. Prepare: Draw a picture of the situation, identify the system, in this case the mountain climber, and draw a closed curve around it. Name and label all relevant contact forces and long-range forces.

Tension $\vec{T}$

Weight $\vec{w}$

Solve: There are two forces acting on the mountain climber due to her interactions with the two agents earth and rope. One of the forces *on* the climber is the long-range weight force *by* the earth. The other force is the tension force exerted *by* the rope.

Assess: Note that there are no horizontal forces.

P4.9. Prepare: Draw a picture of the situation, identify the system, in this case the baseball player, and draw a closed curve around it. Name and label all relevant contact forces and long-range forces.

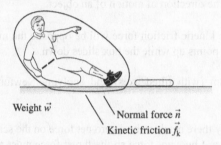

Weight $\vec{w}$

Normal force $\vec{n}$

Kinetic friction $\vec{f}_k$

Solve: There are three forces acting *on* the baseball player due to his interactions with the two agents earth and ground. One of the forces *on* the player is the long-range weight force *by* the earth. Another force is the normal force exerted *by* the ground due to the contact between him and the ground. The third force is the kinetic friction force *by* the ground due to his sliding motion on the ground.

Assess: Note that the kinetic friction force would be *absent* if the baseball player were *not* sliding.

P4.11. Prepare: We follow the outline in Tactics Box 4.2. See also Conceptual Example 4.2.

The exact angle of the slope is not critical in this problem; the answers would be very similar for any angle between 0° and 90°.

Solve: The system is the skier.

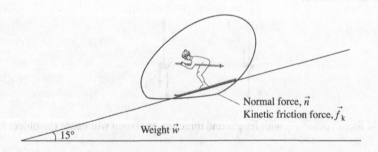

Normal force, $\vec{n}$

Kinetic friction force, $\vec{f}_k$

15°   Weight $\vec{w}$

To identify forces, think of objects that are in contact with the object under consideration, as well as any long-range forces that might be acting on it. We are told to not ignore friction, but we will ignore air resistance.

The objects that are in contact with the skier are the snow-covered slope and. . . and that's all (although we will identify two forces exerted by this agent). The long-range force on the skier is the gravitational force of the earth on the skier.

One of the forces, then, is the gravitational force of the earth on the skier. This force points straight toward the center of the earth.

The slope, as we mentioned, exerts two forces on the skier: the normal force (directed perpendicularly to the slope) and the frictional force (directed parallel to the slope, backward from the downhill motion).

Assess: Since there are no other objects (agents) in contact with the skier (we are ignoring the air, remember?) and no other long-range forces we can identify (the gravitational force of the moon or the sun on the skier is also too small to be worth mentioning), then we have probably catalogued them all.

We are not told whether the skier has a constant velocity or is accelerating, and that factor would influence the relative lengths of the three arrows representing the forces. If the motion is constant velocity, then the vector sum of the three arrows must be zero.

P4.13. Prepare: Refer to Figure P4.13. From force = mass × acceleration or mass = force/acceleration or mass = 1/(acceleration/force), mass is

$$m = \frac{1}{\text{slope of the acceleration-versus-force graph}}$$

A larger slope implies a smaller mass.

Solve: We know $m_2 = 0.20$ kg, and we can find the other masses relative to m_2 by comparing their slopes. Thus

$$\frac{m_1}{m_2} = \frac{1/\text{slope 1}}{1/\text{slope 2}} = \frac{\text{slope 2}}{\text{slope 1}} = \frac{1}{5/2} = \frac{2}{5} = 0.40$$

$$\Rightarrow m_1 = 0.40\, m_2 = 0.40 \times 0.20\,\text{kg} = 0.080\,\text{kg}$$

Similarly,

$$\frac{m_3}{m_2} = \frac{1/\text{slope 3}}{1/\text{slope 2}} = \frac{\text{slope 2}}{\text{slope 3}} = \frac{1}{2/5} = \frac{5}{2} = 2.50$$

$$\Rightarrow m_3 = 2.50\, m_2 = 2.50 \times 0.20\,\text{kg} = 0.50\,\text{kg}$$

Assess: From the initial analysis of the slopes, we had expected $m_3 > m_2$ and $m_1 < m_2$. This is consistent with our numerical answers.

P4.17. Prepare: The problem may be solved by applying Newton's second law to the present and the new situation.
Solve: (a) We are told that for an unknown force (call it F_o) acting on an unknown mass (call it m_o) the acceleration of the mass is 8.0 m/s². According to Newton's second law

$$F_o = m_o (8.0\ \text{m/s}^2) \quad \text{or} \quad F_o/m_o = 8.0\ \text{m/s}^2$$

For the new situation, the new force is $F_{new} = 2F_o$, the mass is not changed $(m_{new} = m_o)$ and we may find the acceleration by

$$F_{new} = m_{new} a_{new}$$

or

$$a_{new} = F_{new}/m_{new} = 2F_o/m_o = 2(F_o/m_o) = 2(8\ \text{m/s}^2) = 16\ \text{m/s}^2$$

(b) For the new situation, the force is unchanged $F_{new} = F_o$, the new mass is half the old mass $(m_{new} = m_o/2)$ and we may find the acceleration by

$$F_{new} = m_{new} a_{new}$$

or

$$a_{new} = F_{new}/m_{new} = F_o/2m_o = (F_o/m_o)/2 = (8.0\ \text{m/s}^2)/2 = 4.0\ \text{m/s}^2$$

(c) A similar procedure gives $a = 8.0$ m/s².

(d) A similar procedure gives $a = 32$ m/s².

Assess: From the algebraic relationship $a = F/m$ we can see that when (a) the force is doubled, the acceleration is doubled; (b) the mass is doubled, the acceleration is halved; (c) both force and mass are doubled, the acceleration doesn't change; and (d) force is doubled and mass is halved, the acceleration will be four times larger.

P4.21. Prepare: The graph shows acceleration vs. force.
Solve: Newton's second law is $F = ma$. We can read a force and an acceleration from the graph, and hence find the mass. Choosing the force $F = 1$ N gives us $a = 4$ m/s². Newton's second law then yields $m = 0.25$ kg.

Assess: Slope of the acceleration-versus-force graph is 4 m/N·s², and therefore, the inverse of the slope will give the mass.

P4.23. Prepare: We can use Newton's second law to find the acceleration of the bear.
Solve: The only forces on the bear are exerted by the girl and boy. A free-body diagram is shown.

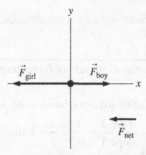

(a) From the free-body diagram shown, the net force in the x-direction is
$$\vec{F}_{Net} = \vec{F}_{Boy} + \vec{F}_{Girl} = 15 \text{ N} - 17 \text{ N} = -2 \text{ N}$$
The net force acting on the bear is 2 N to the left. Since the net force on the bear is not zero, the bear is accelerating. Since at this instant we know nothing about the rate at which the bear's position is changing nothing can be said about the velocity of the bear.
(b) The bear is accelerating, since there is a net force on the bear. From part **(a)**, the net force is 2 N to the left. We can use Newton's second law to find the acceleration of the bear given the mass of the bear.
$$a = \frac{F_{net}}{m} = \frac{-2 \text{ N}}{0.2 \text{ kg}} = -10 \text{ m/s}^2$$
The acceleration is in the same direction as the force, to the left.
Assess: Knowing the mass of an object and the net force acting on it, Newton's second law may be used to determine its acceleration. The acceleration is always in the direction of the net force acting on an object.

P4.27. Prepare: The free-body diagram shows two equal and opposite forces such that the net force is zero. The force directed down is labeled as a weight, and the force directed up is labeled as a tension. With zero net force the acceleration is zero. Draw as shown a picture of a real object with two forces to match the given free-body diagram.

Solve: A possible description is: "An object hangs from a rope and is at rest." Or, "An object hanging from a rope is moving up or down with a constant speed."
Assess: This problem and the following two problems make it clear how important it is to know all forces (and their direction) acting on an object in order to determine the net force acting on the object.

P4.29. Prepare: The free-body diagram shows three forces. There is a weight force $\vec{w}$, which is down. There is a normal force labeled $\vec{n}$, which is up. The forces $\vec{w}$ and $\vec{n}$ are shown with vectors of the same length so they are equal in magnitude and the net vertical force is zero. So we have an object on the ground that is not moving vertically. There is also a force $\vec{f}_k$ to the left. This must be a frictional force and we need to decide whether it is static or kinetic friction. The frictional force is the only horizontal force, so the net horizontal force must be $\vec{f}_k$. This means there is a net force to the left producing an acceleration to the left. This all implies motion and therefore the frictional force is kinetic. Draw a picture of a real object with three forces to match the given free-body diagram.

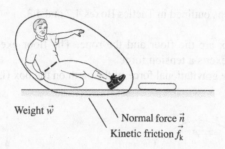

Weight $\vec{w}$

Normal force $\vec{n}$
Kinetic friction $\vec{f}_k$

Solve: A possible description is, "A baseball player is sliding into second base."
Assess: On the free-body diagram, kinetic friction force is the only horizontal force, and it is pointing to the left. This tells us that the baseball player is sliding to the right.

P4.31. Prepare: We will follow the procedures in Tactics Box 4.2 and Tactics Box 4.3.
Solve: Your car is the system. See the following diagram.

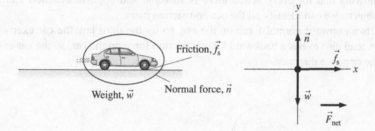

Friction, $\vec{f}_s$

Weight, $\vec{w}$ Normal force, $\vec{n}$

There are contact forces where the car touches the road. One of them is the normal force of the road on the car. The other is the force of static friction between the car's tires and the road, since the car is accelerating from a stop. The only long-range force acting is the weight of the car. Compare to Figure 4.30 and the discussion of propulsion in the text.
Assess: Tactics Box 4.2 and Tactics Box 4.3 give a systematic method for determining all forces on an object and drawing a free-body diagram.

P4.35. Prepare: Follow the steps outlined in Tactics Boxes 4.2 and 4.3. Draw a picture of the situation, identify the system, in this case the car, and draw a closed curve around it. Name and label all relevant contact forces (the normal force and the drag of the air) and long-range forces (weight). Since the road is steep we will assume the car accelerates down the hill; this affects the relative lengths of the arrows we draw.

Solve:

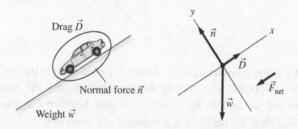

Drag $\vec{D}$

Normal force $\vec{n}$

Weight $\vec{w}$

There are three forces acting *on* the car due to its interactions with the two agents earth and the cable. One of the forces *on* the elevator is the long-range weight force *by* the earth. Another force is the normal force of the road on the car. The third is the drag force of the air on the car.

Assess: We ignored friction but not the air resistance and came out with a reasonable answer.

P4.37. Prepare: We follow the steps outlined in Tactics Boxes 4.2 and 4.3.
Solve: The system is the box.
The objects in contact with the box are the floor and the rope. The floor exerts an upward normal force and a backwards friction force. The rope exerts a tension force.
The important long-range force is the gravitational force of the earth on the box (i.e., the weight).

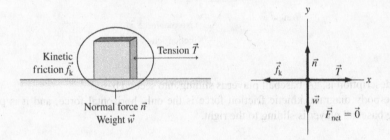

Assess: The net force is zero, as it should be for an object which is moving at constant velocity.

P4.41. Prepare: Knowing that for every action there is an equal and opposite reaction and that these forces are exerted on different objects, we can identify all the action-reaction pairs.
Solve: The road exerts an upward normal force on the car, so by the third law the car exerts a downward normal force on the road. The road also exerts a backward kinetic friction force on the car, so the car exerts a kinetic friction force on the road in the opposite direction.

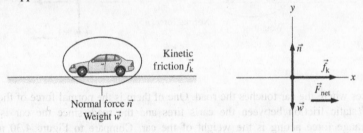

Assess: The road exerts two forces on the car, and the earth exerts a third force on the car; these are the forces that appear in the free-body diagram.

P4.45. Prepare: Redraw the motion diagram as shown.

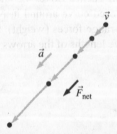

Solve: The velocity vector in the previous figure is shown downward and to the left. So movement is downward and to the left. The velocity vectors get successively longer which means the speed is increasing. Therefore the acceleration is downward and to the left. By Newton's second law $\vec{F} = m\vec{a}$, the net force must be in the same direction as the acceleration. Thus, the net force is downward and to the left.

Assess: Since the object is speeding up, the acceleration vector must be parallel to the velocity vector. This means the acceleration vector must be pointing along the direction of velocity. Therefore the net force must also be downward and to the left.

P4.47. Prepare: Refer to Tactics Box 4.2 and Tactics Box 4.3 for identification of forces and for drawing free-body diagrams. We will draw a correct free-body diagram and compare.

Solve: Your car is the system. See the following diagram.

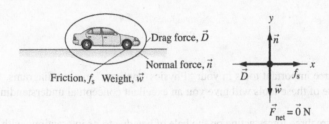

There are contact forces where the car touches the road. One of them is the normal force of the road on the car. The other is the force of static friction between the car's tires and the road since the car is moving.

In addition to this there must be a force in the opposite direction to the car's motion, since the car is moving at constant speed. If only the frictional force acted in the horizontal direction, the car would be accelerating! The diagram omits one of the forces. A possible force that acts in this direction is the force of air drag on the car, which is indicated on the diagram.

The only long-range force acting is the weight of the car.

The diagram also identifies the weight of the car and the normal force on the car as an action/reaction pair. This isn't possible, since both these forces act on the same object, while action/reaction pairs always act on *different* objects. The normal force on an object and its weight are *never* action/reaction pairs.

Assess: In order for an object to be moving at constant velocity, the net force on it must be zero. Action/reaction pairs always act on two different objects.

P4.51. Prepare: The normal force is perpendicular to the ground. The thrust force is parallel to the ground and in the direction of acceleration. The drag force is opposite to the direction of motion. There are four forces acting *on* the jet plane due to its interactions with the four agents the earth, the air, the ground, and the hot gases exhausted to the environment. One force on the rocket is the long-range weight force *by* the earth. The second force is the drag force *by* the air. Third is the normal force on the rocket *by* the ground. The fourth is the thrust force exerted on the jet plane *by* the hot gas that is being let out to the environment. Since the jet plane is speeding down the runway, its acceleration is pointing to the right. Therefore, the net force on the jet plane must also point to the right.

Now, draw a picture of the situation, identify the system, in this case the jet plane, and draw a motion diagram. Draw a closed curve around the system, and name and label all relevant contact forces and long-range forces.

Solve: A force-identification diagram, a motion diagram, and a free-body diagram are shown.

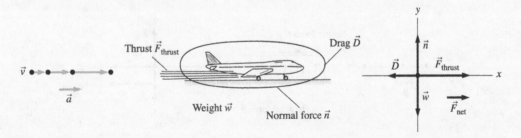

Assess: You now have three important tools in your "Physics Toolbox," motion diagrams, force diagrams, and free-body diagrams. Careful use of these tools will give you an excellent conceptual understanding of a situation.

P4.53. Prepare: The kinetic friction force is what propels her down the track.
Solve: A force-identification diagram, a motion diagram, and a free-body diagram are shown.

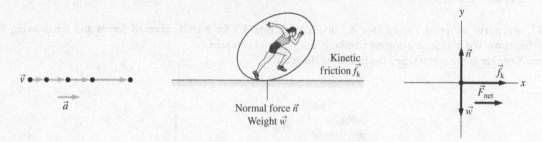

Assess: You now have three important tools in your "Physics Toolbox," motion diagrams, force diagrams, and free-body diagrams. Careful use of these tools will give you an excellent conceptual understanding of a situation.

P4.55. Prepare: There are three forces acting on the bale of hay due to its interactions with the two agents: the earth and the bed of the truck. The two contact forces between the bale of hay and the bed of the truck are the normal force and the force of kinetic friction which is dragging the bale of hay forward (even though it is sliding backward). The force at a distance is the force the earth exerts on the bale of hay (the weight). Since the normal force and the weight are equal in magnitude and opposite in direction, there is no net vertical force. Since the force of kinetic friction provides a net horizontal force, the net force acting on the bale of hay and hence the acceleration of the bale of hay is in the direction of the force of kinetic friction. Now, draw a picture of the situation, identify the system, in this case the bale of hay, and draw a motion diagram. Draw a closed curve around the system, and name and label all relevant contact forces and long-range forces.
Solve:

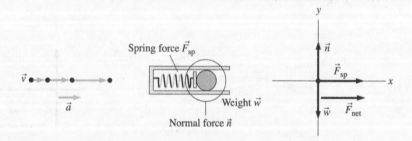

Assess: Since there is a net or unbalanced force acting on the bale of hay, it will experience an acceleration in the direction of this force.

P4.57. Prepare: The ball rests on the floor of the barrel because the weight is equal to the normal force. There is a force of the spring to the right, which causes acceleration. Now, draw a picture of the situation, identify the system, in this case the plastic ball, and draw a motion diagram. Draw a closed curve around the system, and name and label all relevant contact forces and long-range forces. Neglect friction
Solve: A force-identification diagram, a motion diagram, and a free-body diagram are shown.

Assess: Since the normal force acting on the ball and the weight of the ball are equal in magnitude and opposite in direction, the ball experiences no vertical motion.

P4.61. Prepare: You can see from the motion diagram that the bag accelerates to the left along with the car as the car slows down. According to Newton's second law, $\vec{F} = m\vec{a}$, there must be a force to the *left* acting on the bag. This is friction, but not kinetic friction. The bag is not sliding across the seat. Instead, it is static friction, the force that prevents slipping. Were it not for static friction, the bag would slide off the seat as the car stops. Static friction acts in the direction needed to prevent slipping. In this case, friction must act in the backward (toward the left) direction.

Now, draw a picture of the situation, identify the system, in this case the bag of groceries, and draw a motion diagram. Draw a closed curve around the system, and name and label all relevant contact forces and long-range forces.

Solve: A force-identification diagram, a motion diagram, and a free-body diagram are shown.

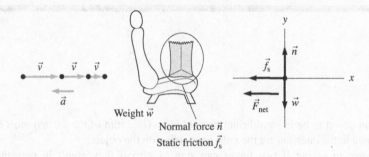

Assess: Since the normal force acting on the bag of groceries and the weight of the groceries are equal in magnitude and opposite in direction, the bag experiences no vertical motion. The only horizontal force acting on the bag of groceries is static friction, and it provides the net force acting on the bag which results in the acceleration of the bag.

P4.65. Prepare: Assume the ball undergoes constant acceleration during the pitch so we can use the kinematic equations.

$$(v_x)_f^2 = (v_x)_i^2 + 2a_x \Delta x$$

Use coordinates where $+x$ is in the direction the ball is thrown. We are given $\Delta x = 1.0$ m, $(v_x)_f = 46$ m/s, and $m = 0.145$ kg. Assume $(v_x)_i = 0.0$ m/s.

We'll first solve for a_x and then use Newton's second law to find the average force.

Solve: (a)

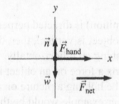

(b) Solve the equation for a_x.

$$a_x = \frac{(v_x)_f^2 - (v_x)_i^2}{2\Delta x} = \frac{(46 \text{ m/s})^2 - (0.0 \text{ m/s})^2}{2(1.0 \text{ m})} = 1060 \text{ m/s}^2$$

$$F_x = ma_x = (0.145 \text{ kg})(1060 \text{ m/s}^2) = 150 \text{ N}$$

(c) Say a typical pitcher weighs 170 lbs.

$$170 \text{ lb}\left(\frac{4.45 \text{ N}}{1 \text{ lb}}\right) \approx 760 \text{ N}$$

Now divide the force from part (b) by this weight to see the fraction.

$$150 \text{ N} \div 760 \text{ N} \approx \frac{1}{5}$$

So the force the pitcher exerted on the ball is about 1/5 his weight.

Assess: The answer to each part seems reasonable. The units also work out.

5

APPLYING NEWTON'S LAWS

Q5.1. Reason: For an object to be in equilibrium, the net force (i.e., sum of the forces) must be zero. Assume that the two forces mentioned in the question are the only ones acting on the object.

The question boils down to asking if two forces can sum to zero if they aren't in opposite directions. Mental visualization shows that the answer is no, but so does a careful analysis. Set up a coordinate system with the x-axis along one of the forces. If the other force is not along the negative x-axis then there will be a y (or z) component that cannot be canceled by the first one along the x-axis.

Assess: In summary, two forces not in opposite directions cannot sum to zero. Neither can two forces with different magnitudes. However, three can.

Q5.3. Reason: Assume you are sitting in a chair, and that you are at rest. (Parts of your body may be moving, but if you model your body as a particle, then you generally aren't moving much as you read.)

The two forces that act on you are the gravitational force of the earth on you directed down and the normal force of the chair pushing up on you. These two forces are equal in magnitude and opposite in direction and so the sum (the net force) is zero.

Assess: When you aren't accelerating Newton's second law says you experience a zero net force. So this analysis would apply equally to the case of you reading this while sitting in a chair on a smoothly moving (constant velocity) train.

Q5.7. Reason: The normal force (by definition) is directed perpendicular to the surface.

(a) If the surface that exerts a force on an object is vertical, then the normal force would be horizontal. An example would be holding a picture on a wall by pushing on it horizontally. The wall would exert a normal force horizontally.

(b) In a similar vein, if the surface that exerts a force on an object is horizontal and above the object, then the normal force would be down. One example would be holding a picture on a ceiling by pushing on it. The ceiling would exert a normal force vertically downward. Another example would be the Newton's third law pair force in the case of you sitting on a chair; the chair exerts a normal force upward on you, so you exert a normal force downward on the chair.

Assess: We see that the normal force can be in any direction; it is always perpendicular to the surface pushing on the object in question.

Q5.9. Reason: Increasing the mass does increase the net force on the system, but it also increases the inertia. $a = \dfrac{F_{net}}{m}$. Since both the net force and mass are increased they still cancel, leaving the acceleration the same.

Assess: The m cancels out of every term.

Q5.13. Reason: The skydiver is falling with a constant velocity just before she opens her parachute. At this point, the drag force on the diver is equal to her weight. When she opens her parachute, her effective area is increased, so this increases the drag force on the diver. This will cause a net force on the diver in the upward direction, which is greater than her weight and will decelerate her.

Assess: This makes sense. The parachute slows the diver down further.

Q5.15. Reason: As the plane's thrust is decreased, the plane will start to decelerate. From Equation 5.12, as the plane's velocity decreases, so does the drag force on the plane. The plane's velocity will decrease until the drag force equals the thrust force. At that point, the plane will stop decelerating as it reaches a new equilibrium. If the drag force continued decreasing, the thrust force would re-accelerate the plane to the point where it would stop accelerating, again reaching the same equilibrium. Therefore the plane will travel with a constant velocity once the new equilibrium is reached.
Assess: Drag force decreases with decrease in velocity.

Q5.21. Reason: The kinetic friction acts in a direction to oppose the relative motion, so on block 1 the kinetic friction is to the right and on block 2 it is to the left.
Assess: We would expect them to be opposite since they are a Newton third law pair and the forces in a third law pair are always in opposite directions.

Q5.23. Reason: The ball is in equilibrium. We will use Equation 5.1.
See the free-body diagram below.

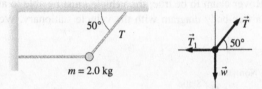

In the vertical direction we have

$$T\sin(50°) - w = T\sin(50°) - mg = 0$$

Solving for T, we obtain

$$T = \frac{mg}{\sin(50°)} = \frac{(2.0\ \text{kg})(9.80\ \text{m/s}^2)}{\sin(50°)} = 26\ \text{N}$$

The correct choice is D.
Assess: Note that we did not need to use the horizontal components of the forces.

Q5.25. Reason: We will use Equation 5.2 since neither the dog nor the floor is in equilibrium.

(a)

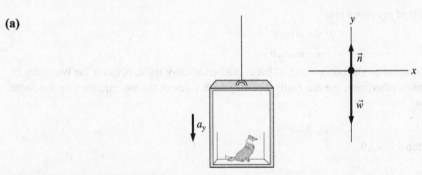

From the free-body diagram above, we have $n - w = ma_y$.

Solving for the normal force,
$$n = w + ma_y = mg + ma_y = (5.0\ \text{kg})(9.80\ \text{m/s}^2) + (5.0\ \text{kg})(-1.20\ \text{m/s}^2) = 43\ \text{N}$$

The correct choice is B.
(b) The normal force on the dog is the force of the floor of the elevator on the dog. The force of the dog on the elevator floor is the reaction force to this. The correct choice is D.
Assess: This result makes sense; the normal force will be less than the weight of the dog, which is 49 N.

Q5.27. Reason: This is still a Newton's second law question; the only twist is that the object is not in equilibrium, i.e., the right side of the second law is not zero.

The forces on Eric are the downward gravitational force of the earth on him w, and the upward normal force of the scale on him n (which we want to know).

We note that $a = -1.7 \text{ m/s}^2$ and $w = mg = (60 \text{ kg})(9.80 \text{ m/s}^2) = 5.88 \text{ N}$.

This is a one-dimensional question in the vertical direction, so the following equations are all in the y-direction.

$$F_{\text{net}} = ma$$

$$n - w = ma$$

$$n = ma + w = (60 \text{ kg})(-1.7 \text{ m/s}^2) + 588 \text{ N} = 486 \text{ N} \approx 500 \text{ N}$$

The correct choice is C.

Assess: Because the elevator is accelerating down, we expect the scale to read a bit less than Eric's normal weight. This is the case.

It is important that neither the question nor the answer specify whether the elevator is moving up or down. The elevator can be accelerating down in two ways: It can be moving up and slowing (such as the end of a trip from a low floor to a high floor), or it can be moving down and gaining speed (such as the beginning of a trip from a high floor to a low floor). The answer is the same in both cases.

Q5.31. Reason: For the Land Rover claim to be true, the vehicle must be able to at least sit on the hill motionless without slipping. So we'll draw a free-body diagram with the vehicle stationary. We use tilted axes with the x-axis running up the slope.

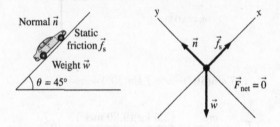

First apply $F_{\text{net}} = ma$ in the y-direction.

$$n - w\cos\theta = 0$$

Then apply $F_{\text{net}} = ma$ in the x-direction.

$$f_s - w\sin\theta = 0$$

With $f_s = \mu_s n$ we rearrange the pair of equations into

$$\mu_s n = w\sin\theta$$

$$n = w\cos\theta$$

Now the key is to divide the top equation by the bottom one. (This is mathematically legal, because the two sides of the bottom equation are equal to each other, then we are really dividing both sides of the top equation by the same thing.) Remember that $\frac{\sin\theta}{\cos\theta} = \tan\theta$.

$$\mu_s = \tan\theta$$

Insert $\theta = 45°$ and we have $\mu_s = \tan 45° = 1.0$.

The correct choice is D.

Assess: The answer to this question is independent of the mass of the Land Rover! An equivalent way to express this is that w (and n) cancelled out.

Also notice that by solving the equations with a variable θ and only inserting the value of 45° at the end, we are able to solve for the required minimum μ_s for any angle.

Problems

P5.1. Prepare: The massless ring is in static equilibrium, so all the forces acting on it must cancel to give a zero net force. The forces acting on the ring are shown on a free-body diagram below.

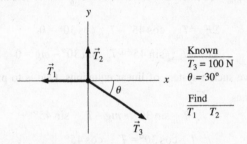

Known
$T_3 = 100$ N
$\theta = 30°$

Find
T_1 T_2

Solve: Written in component form, Newton's first law is

$$(F_{net})_x = \Sigma F_x = T_{1x} + T_{2x} + T_{3x} = 0 \text{ N} \quad (F_{net})_y = \Sigma F_y = T_{1y} + T_{2y} + T_{3y} = 0 \text{ N}$$

Evaluating the components of the force vectors from the free-body diagram:

$$T_{1x} = -T_1 \quad T_{2x} = 0 \text{ N} \quad T_{3x} = T_3 \cos 30°$$
$$T_{1y} = 0 \text{ N} \quad T_{2y} = T_2 \quad T_{3y} = -T_3 \sin 30°$$

Using Newton's first law:

$$-T_1 + T_3 \cos 30° = 0 \text{ N} \quad T_2 - T_3 \sin 30° = 0 \text{ N}$$

Rearranging:

$$T_1 = T_3 \cos 30° = (100 \text{ N})(0.8666) = 87 \text{ N} \quad T_2 = T_3 \sin 30° = (100 \text{ N})(0.5) = 50 \text{ N}$$

Assess: Since $\vec{T}_3$ acts closer to the x-axis than to the y-axis, it makes sense that $T_1 > T_2$.

P5.5. Prepare: The femur is in static equilibrium. We can use Equation 5.1.
Solve: See the free-body diagram below.

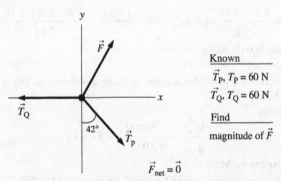

Known
$\vec{T}_P, T_P = 60$ N
$\vec{T}_Q, T_Q = 60$ N

Find
magnitude of $\vec{F}$

The direction of the force the femur exerts on the patella is indicated roughly on the previous diagram. The sum of the x-components of the forces must be zero. This gives

$$T_Q = T_P \sin(42°) + F_x$$

Solving for F_x,

$$F_x = T_Q - T_P \sin(42°) = 60 \text{ N} - (60 \text{ N}) \sin(42°) = 20 \text{ N}$$

The sum of the y-components of the forces must be zero also. This gives

$$F_y = T_P \cos(42°) = (60 \text{ N}) \cos(42°) = 45 \text{ N}$$

The magnitude of the force by the femur on the patella is then

$$F = \sqrt{(F_x)^2 + (F_y)^2} = \sqrt{(20 \text{ N})^2 + (45 \text{ N})^2} = 49 \text{ N}$$

Assess: This result is reasonable in magnitude, considering the magnitude of the forces exerted by the tendons and their directions.

P5.7. Prepare: The tension in the more vertical of the two angled ropes (the right one) will have a greater tension, so we apply Newton's second law and set $T_{right} = 1500$ N and solve for m. T_{left} will be less than 1500 N and will not break.
Solve:

$$\Sigma F_x = T_{right} \cos 45° - T_{left} \cos 30° = 0$$

$$\Sigma F_y = T_{right} \sin 45° + T_{left} \sin 30° - mg = 0$$

There are various strategies to solve such a system of linear equations. One is to put the two T_{left} terms on the left side and then divide the two equations.

$$T_{left} \sin 30° = mg - T_{right} \sin 45°$$

$$T_{left} \cos 30° = T_{right} \cos 45°$$

Now dividing these two equations cancels T_{left} on the left (since we don't need T_{left}) and leaves $\tan 30°$.

$$\tan 30° = \frac{mg - T_{right} \sin 45°}{T_{right} \cos 45°}$$

Solve for m and set $T_{right} = 1500$ N.

$$m = \frac{T_{right}(\tan 30° \cos 45° + \sin 45°)}{g} = \frac{(1500 \text{ N})(\tan 30° \cos 45° + \sin 45°)}{9.80 \text{ m/s}^2} = 170 \text{ kg}$$

Assess: The answer seems reasonable, since if there were only one vertical rope it could hold $(1500 \text{ N})/(9.80 \text{ m/s}^2) = 153$ kg and here we have the left rope to help.

P5.11. Prepare: The free-body diagram shows five forces acting on an object whose mass is 2.0 kg. All the forces point along x- or y-axes. We will first find the net force along the x- and the y-axes and then divide these forces by the object's mass to obtain the x- and y-components of the object's acceleration.
Solve: Applying Newton's second law:

$$a_x = \frac{(F_{net})_x}{m} = \frac{4 \text{ N} - 2 \text{ N}}{2 \text{ kg}} = 1.0 \text{ m/s}^2 \qquad a_y = \frac{(F_{net})_y}{m} = \frac{3 \text{ N} - 1 \text{ N} - 2 \text{ N}}{2 \text{ kg}} = 0.0 \text{ m/s}^2$$

Assess: The object's acceleration is only along the x-axis.

P5.15. Prepare: We must first find the astronaut's mass on earth and then multiply it with Mars's acceleration due to gravity to find his weight on Mars.
Solve: The mass of the astronaut is

$$m = \frac{w_{earth}}{g_{earth}} = \frac{800 \text{ N}}{9.80 \text{ m/s}^2} = 81.6 \text{ kg}$$

Therefore, the weight of the astronaut on Mars is

$$w_{Mars} = mg_{Mars} = (81.6 \text{ kg})(3.76 \text{ m/s}^2) = 310 \text{ N}$$

Assess: The smaller acceleration of gravity on Mars reveals that objects are less strongly attracted to Mars than to the earth, so the smaller weight on Mars makes sense. Also, note that the astronaut's mass stays unchanged.

P5.17. Prepare: The true weight of an object and its apparent weight are connected by $w_{app} = m(g + a_y)$. After the box, with a passenger inside, leaves the rubber band (still moving upward), the box as well as the passenger are falling freely and their acceleration is equal to $-g$.

Solve: The passenger's apparent weight will be $(75 \text{ kg})(9.80 \text{ m/s}^2) + (-9.80 \text{ m/s}^2) = 0$.

Assess: The apparent weight is always zero for objects in free fall.

P5.21. Prepare: We'll assume Zach is a particle moving under the effect of two forces acting in a single vertical line: gravity and the supporting force of the elevator. These forces are shown in Figure 5.9 in a free-body diagram.

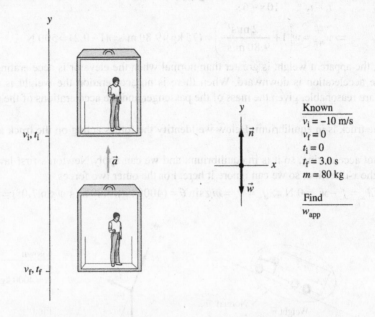

Known
$v_i = -10$ m/s
$v_f = 0$
$t_i = 0$
$t_f = 3.0$ s
$m = 80$ kg

Find
w_{app}

Solve: (a) Before the elevator starts braking, Zach is not accelerating. His apparent weight is

$$w_{app} = w\left(1 + \frac{a}{g}\right) = w\left(1 + \frac{0 \text{ m/s}^2}{g}\right) = mg = (80 \text{ kg})(9.80 \text{ m/s}^2) = 784 \text{ N}$$

or 780 N to two significant figures.
(b) Using the definition of acceleration,

$$a = \frac{\Delta v}{\Delta t} = \frac{v_f - v_i}{t_f - t_i} = \frac{0 - (-10) \text{ m/s}}{3.0 \text{ s}} = 3.33 \text{ m/s}^2$$

$$\Rightarrow w_{app} = w\left(1 + \frac{a}{g}\right) = (80 \text{ kg})(9.80 \text{ m/s}^2)\left(1 + \frac{3.33 \text{ m/s}^2}{9.80 \text{ m/s}^2}\right) = (784 \text{ N})(1 + 0.340) = 1100 \text{ N}$$

Assess: While the elevator is braking, it not only must support Zach's weight but must also push upward on him to decelerate him, so the apparent weight is greater than his normal weight.

P5.23. Prepare: The passenger is acted on by only two vertical forces: the downward pull of gravity and the upward force of the elevator floor. Referring to Figure P5.23, the graph has three segments corresponding to different conditions: (1) increasing velocity, meaning an upward acceleration, (2) a period of constant upward velocity, and (3) decreasing velocity, indicating a period of deceleration (negative acceleration). Given the assumptions of our model, we can calculate the acceleration for each segment of the graph.
Solve: The acceleration for the first segment is

$$a_y = \frac{v_f - v_i}{t_f - t_i} = \frac{8 \text{ m/s} - 0 \text{ m/s}}{2 \text{ s} - 0 \text{ s}} = 4 \text{ m/s}^2 \Rightarrow w_{app} = w\left(1 + \frac{a_y}{g}\right) = (mg)\left(1 + \frac{4 \text{ m/s}^2}{9.80 \text{ m/s}^2}\right)$$

$$= (75 \text{ kg})(9.80 \text{ m/s}^2)\left(1 + \frac{4}{9.8}\right) = 1000 \text{ N}$$

For the second segment, $a_y = 0$ m/s^2 and the apparent weight is

$$w_{app} = w\left(1 + \frac{0 \text{ m/s}^2}{g}\right) = mg = (75 \text{ kg})(9.80 \text{ m/s}^2) = 740 \text{ N}$$

For the third segment,

$$a_y = \frac{v_3 - v_2}{t_3 - t_2} = \frac{0 \text{ m/s} - 8 \text{ m/s}}{10 \text{ s} - 6 \text{ s}} = -2 \text{ m/s}^2$$

$$\Rightarrow w_{app} = w\left(1 + \frac{-2 \text{ m/s}^2}{9.80 \text{ m/s}^2}\right) = (75 \text{ kg})(9.80 \text{ m/s}^2)(1 - 0.2) = 590 \text{ N}$$

Assess: As expected, the apparent weight is greater than normal when the elevator is accelerating upward and lower than normal when the acceleration is downward. When there is no acceleration the weight is normal. In all three cases the magnitudes are reasonable, given the mass of the passenger and the accelerations of the elevator.

P5.27. Prepare: The truck is in equilibrium. Below we identify the forces acting on the truck and construct a free-body diagram.
Solve: The truck is not accelerating, so it is in equilibrium, and we can apply Newton's first law. The normal force has no component in the x-direction, so we can ignore it here. For the other two forces

$$(F_{net})_x = \Sigma F_x = f_s - w_x = 0 \text{ N} \Rightarrow f_s = w_x = mg\sin\theta = (4000 \text{ kg})(9.80 \text{ m/s}^2)(\sin 7.0°) = 4800 \text{ N}$$

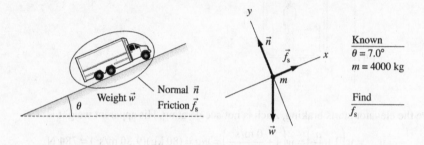

Assess: The truck's weight (mg) is roughly 40,000 N. A friction force that is $\approx$12% of the truck's weight seems reasonable.

P5.29. Prepare: The pig is acted on by two opposing forces in a single line: the farmer's pull and the friction. The pig will be subject to static friction until (and if!) it begins to move; after that it will be subject to kinetic friction. We give below an overview of the pictorial representation, a free-body diagram, and a list of values. We will calculate the force of maximum static friction and compare it with the maximum applied force.

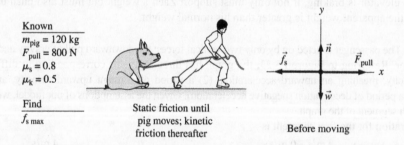

Solve: Since the pig does not accelerate in the vertical direction, the free-body diagram shows that $n = w = mg$. The maximum friction force is

$$f_{s\,max} = \mu_s mg = (0.8)(120 \text{ kg})(9.80 \text{ m/s}^2) = 940 \text{ N}$$

The maximum static friction force is greater than the farmer's maximum pull of 800 N; thus, the farmer will not be able to budge the pig.
Assess: The farmer should have known better.

P5.33. Prepare: We will need to apply Newton's second law in both the vertical and horizontal directions. We want to use the coefficient of static friction since we want the box to stay stationary.

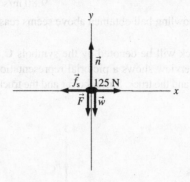

Solve:

$$\Sigma F_y = n - F - mg = 0 \Rightarrow n = F + mg$$
$$\Sigma F_x = 125\,\text{N} - f_s = 125\,\text{N} - \mu_s n = 125\,\text{N} - \mu_s(F + mg) = 0$$

Solve for F.

$$F = \frac{125\,\text{N}}{\mu_s} - mg = \frac{125\,\text{N}}{0.35} - (30\,\text{kg})(9.80\,\text{m/s}^2) = 63\,\text{N}$$

Assess: 63 N is about half of the force on the rope; this seems reasonable given μ_s.

P5.35. Prepare: The bowling ball falls straight down toward the earth's surface. The bowling ball is subject to a net force that is the resultant of the weight and drag force vectors acting vertically in the downward and upward directions, respectively. Once the net force acting on the ball becomes zero, the terminal velocity is reached and remains constant for the rest of the motion. An overview of a pictorial representation and a free-body diagram are shown later.

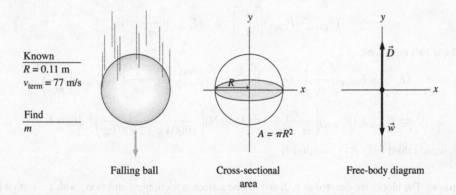

Falling ball   Cross-sectional   Free-body diagram
                              area

Solve: The mathematical equation defining the dynamical equilibrium situation for the falling ball is

$$\vec{F}_{net} = \vec{w} + \vec{D} = \vec{0}\,\text{N}$$

Since only the vertical direction matters, one can write

$$\Sigma F_y = 0\,\text{N} \Rightarrow F_{net} = D - w = 0\,\text{N}$$

When this condition is satisfied, the speed of the ball becomes the constant terminal speed $v = v_{term}$. The magnitudes of the weight and drag forces acting on the ball are

$$w = mg = m(9.80\,\text{m/s}^2)$$

$$D \approx \frac{1}{4}\rho\left(A v_{term}^2\right) = 0.25\rho(\pi R^2)v_{term}^2 = (0.25\pi)(1.22\,\text{kg/m}^3)(0.11\,\text{m})^2(77\,\text{m/s})^2 = 68.7\,\text{N}$$

The condition for dynamic equilibrium becomes

$$(9.80 \text{ m/s}^2)m - 68.7 \text{ N} = 0 \text{ N} \Rightarrow m = \frac{68.7 \text{ N}}{9.80 \text{ m/s}^2} = 7.0 \text{ kg}$$

Assess: The value of the mass of the bowling ball obtained above seems reasonable; it's about 19 pounds.

P5.39. Prepare: The car and the truck will be denoted by the symbols C and T, respectively. The ground will be denoted by the symbol G. A visual overview shows a pictorial representation, a list of known and unknown values, and a free-body diagram for both the car and the truck. Since the car and the truck move together in the positive x-direction, they have the same acceleration.

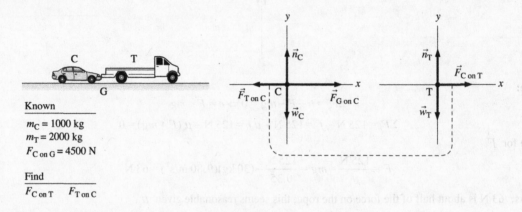

Known

$m_C = 1000 \text{ kg}$
$m_T = 2000 \text{ kg}$
$F_{C \text{ on } G} = 4500 \text{ N}$

Find

$F_{C \text{ on } T}$ $F_{T \text{ on } C}$

Solve: **(a)** The x-component of Newton's second law for the car is

$$\sum(F_{\text{on C}})_x = F_{G \text{ on C}} - F_{T \text{ on C}} = m_C a_C$$

The x-component of Newton's second law for the truck is

$$\sum(F_{\text{on T}})_x = F_{C \text{ on T}} = m_T a_T$$

Using $a_C = a_T = a$ and $F_{T \text{ on } C} = F_{C \text{ on } T}$, we get

$$(F_{C \text{ on } G} - F_{C \text{ on } T})\left(\frac{1}{m_C}\right) = a \quad (F_{C \text{ on } T})\left(\frac{1}{m_T}\right) = a$$

Combining these two equations,

$$(F_{C \text{ on } G} - F_{C \text{ on } T})\left(\frac{1}{m_C}\right) = (F_{C \text{ on } T})\left(\frac{1}{m_T}\right) \Rightarrow F_{C \text{ on } T}\left(\frac{1}{m_C} + \frac{1}{m_T}\right) = (F_{C \text{ on } G})\left(\frac{1}{m_C}\right)$$

$$\Rightarrow F_{C \text{ on } T} = (F_{C \text{ on } G})\left(\frac{m_T}{m_C + m_T}\right) = (4500 \text{ N})\left(\frac{2000 \text{ kg}}{1000 \text{ kg} + 2000 \text{ kg}}\right) = 3000 \text{ N}$$

(b) Due to Newton's third law, $F_{T \text{ on } C} = 3000 \text{ N}$.

P5.41. Prepare: The blocks are denoted as 1, 2, and 3. The surface is frictionless and along with the earth it is a part of the environment. The three blocks are our three systems of interest. The force applied on block 1 is $F_{A \text{ on } 1} = 12 \text{ N}$. The acceleration for all the blocks is the same and is denoted by a. A visual overview shows a pictorial representation, a list of known and unknown values, and a free-body diagram for the three blocks.

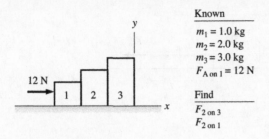

Known
$m_1 = 1.0 \text{ kg}$
$m_2 = 2.0 \text{ kg}$
$m_3 = 3.0 \text{ kg}$
$F_{\text{A on 1}} = 12 \text{ N}$

Find
$F_{2 \text{ on } 3}$
$F_{2 \text{ on } 1}$

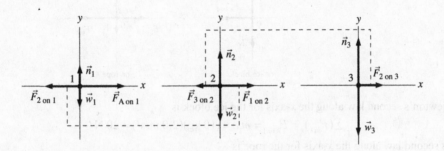

Solve: Newton's second law for the three blocks along the x-direction is

$$\sum(F_{\text{on }1})_x = F_{\text{A on }1} - F_{2 \text{ on }1} = m_1 a \quad \sum(F_{\text{on }2})_x = F_{1 \text{ on }2} - F_{3 \text{ on }2} = m_2 a \quad \sum(F_{\text{on }3})_x = F_{2 \text{ on }3} = m_3 a$$

Adding these three equations and using Newton's third law ($F_{2 \text{ on }1} = F_{1 \text{ on }2}$ and $F_{3 \text{ on }2} = F_{2 \text{ on }3}$), we get

$$F_{\text{A on }1} = (m_1 + m_2 + m_3)a \Rightarrow (12 \text{ N}) = (1.0 \text{ kg} + 2.0 \text{ kg} + 3.0 \text{ kg})a \Rightarrow a = 2.0 \text{ m/s}^2$$

(a) Using this value of a, the force equation on block 3 gives

$$F_{2 \text{ on }3} = m_3 a = (3.0 \text{ kg})(2.0 \text{ m/s}^2) = 6.0 \text{ N}$$

(b) Substituting into the force equation on block 1,

$$12 \text{ N} - F_{2 \text{ on }1} = 12 \text{ N} - (1.0 \text{ kg})(2.0 \text{ m/s}^2) \Rightarrow F_{2 \text{ on }1} = 10 \text{ N}$$

Assess: Because all three blocks are pushed forward by a force of 12 N, the value of 10 N for the force that the 2.0 kg block exerts on the 1.0 kg block is reasonable.

P5.43. Prepare: A visual overview shows below a pictorial representation, a list of known and unknown values, and a free-body diagram for both the ice (I) and the rope (R). The force $\vec{F}_{\text{ext}}$ acts only on the rope. Since the rope and the ice block move together, they have the same acceleration. Also because the rope has mass, F_{ext} on the front end of the rope is not the same as $F_{\text{I on R}}$ that acts on the rear end of the rope.

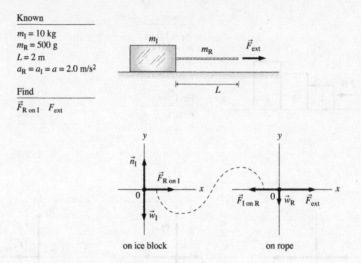

Known

$m_1 = 10$ kg
$m_R = 500$ g
$L = 2$ m
$a_R = a_1 = a = 2.0$ m/s^2

Find

$\vec{F}_{R \text{ on } 1}$ F_{ext}

on ice block   on rope

Solve: **(a)** Newton's second law along the x-axis for the ice block is

$$\Sigma(F_{\text{on } 1})_x = F_{R \text{ on } 1} = m_1 a = (10 \text{ kg})(2.0 \text{ m/s}^2) = 20 \text{ N}$$

(b) Newton's second law along the x-axis for the rope is

$$\Sigma(F_{\text{on } R})_x = F_{ext} - F_{1 \text{ on } R} = m_R a \Rightarrow F_{ext} - F_{R \text{ on } 1} = m_R a \Rightarrow F_{ext} = F_{R \text{ on } 1} + m_R a = 20 \text{ N} + (0.5 \text{ kg})(2.0 \text{ m/s}^2) = 21 \text{ N}$$

Assess: We see that the massless rope approximation is really an aproximation that may not always be good.

P5.47. Prepare: Because the piano is to descend at a steady speed, it is in dynamic equilibrium. The following shows a free-body diagram of the piano and a list of values.

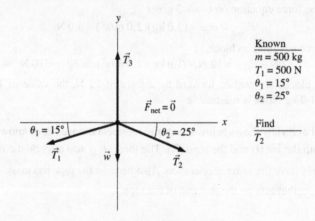

Known

$m = 500$ kg
$T_1 = 500$ N
$\theta_1 = 15°$
$\theta_2 = 25°$

Find

T_2

Solve: **(a)** Based on the free-body diagram, Newton's second law is

$$(F_{net})_x = 0 \text{ N} = T_{1x} + T_{2x} = T_2 \cos\theta - T_1 \cos\theta_1$$

$$(F_{net})_y = 0 \text{ N} = T_{1y} + T_{2y} + T_{3y} + w_y = T_3 - T_1 \sin\theta_1 - T_2 \sin\theta_2 - mg$$

Notice how the force components all appear in the second law with *plus* signs because we are *adding* vector forces. The negative signs appear only when we *evaluate* the various components. These are two simultaneous equations in the two unknowns T_2 and T_3. From the x-equation we find

$$T_2 = \frac{T_1 \cos\theta_1}{\cos\theta_2} = \frac{(500 \text{ N})\cos 15°}{\cos 25°} = 530 \text{ N}$$

(b) Now we can use the y-equation to find

$$T_3 = T_1 \sin\theta_1 + T_2 \sin\theta_2 + mg = 5300 \text{ N}$$

P5.49. Prepare: To find the net force at a given time, we need the acceleration at that time. Because the times where we are asked to find the net force fall on distinct slopes of the velocity-versus-time graph, we can use the constant slopes of the three segments of the graph to calculate the three accelerations.

Solve: For t between 0 s and 3 s,

$$a_x = \frac{\Delta v_x}{\Delta t} = \frac{12 \text{ m/s} - 0 \text{ s}}{3 \text{ s}} = 4 \text{ m/s}^2$$

For t between 3 s and 6 s, $\Delta v_x = 0$ m/s, so $a_x = 0$ m/s². For t between 6 s and 8 s,

$$a_x = \frac{\Delta v_x}{\Delta t} = \frac{0 \text{ m/s} - 12 \text{ m/s}}{2 \text{ s}} = -6 \text{ m/s}^2$$

From Newton's second law, at $t = 1$ s we have

$$F_{\text{net}} = ma_x = (2.0 \text{ kg})(4 \text{ m/s}^2) = 8 \text{ N}$$

At $t = 4$ s, $a_x = 0$ m/s², so $F_{\text{net}} = 0$ N.

At $t = 7$ s,

$$F_{\text{net}} = ma_x = (2.0 \text{ kg})(-6.0 \text{ m/s}^2) = -12 \text{ N}$$

Assess: The magnitudes of the forces look reasonable, given the small mass of the object. The positive and negative signs are appropriate for an object first speeding up, then slowing down.

P5.53. Prepare: The box is acted on by two forces: the tension in the rope and the pull of gravity. Both the forces act along the same vertical line which is taken to be the y-axis. The following shows the free-body diagram for the box.

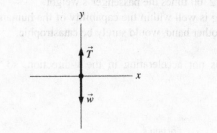

Solve: (a) Since the box is rising at a constant speed, $a_y = 0$ m/s² and the net force on it must be zero:

$$F_{\text{net}} = T - w = 0 \text{ N} \Rightarrow T = w = mg = (50 \text{ kg})(9.80 \text{ m/s}^2) = 490 \text{ N}$$

(b) Since the box is slowing down, $a_y = -5.0$ m/s² and we have

$$F_{\text{net}} = T - w = ma_y = (50 \text{ kg})(-5.0 \text{ m/s}^2) = -250 \text{ N}$$

$$\Rightarrow T = -250 \text{ N} + w = -250 \text{ N} + 490 \text{ N} = 240 \text{ N}$$

Assess: For part **(a)** the zero acceleration immediately implies that the box's weight must be exactly balanced by the upward tension in the rope. For part **(b)**, when the box accelerates downward, the rope need not support the entire weight, hence, T is less than w.

P5.55. Prepare: We can assume the person is moving in a straight line under the influence of the combined decelerating forces of the air bag and seat belt or, in the absence of restraints, the dashboard or windshield. The following is an overview of the situation in a pictorial representation and the occupant's free-body diagram is shown below. Note that the occupant is brought to rest over a distance of 1 m in the former case, but only over 5 mm in the latter.

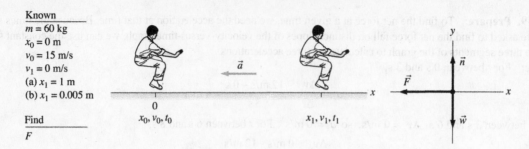

Known
$m = 60$ kg
$x_0 = 0$ m
$v_0 = 15$ m/s
$v_1 = 0$ m/s
(a) $x_1 = 1$ m
(b) $x_1 = 0.005$ m

Find
F

Solve: **(a)** In order to use Newton's second law for the passenger, we'll need the acceleration. Since we don't have the stopping time,

$$v_f^2 = v_i^2 + 2a(x_f - x_i) \Rightarrow a = \frac{v_f^2 = v_i^2}{2(x_f - x_i)} = \frac{0 \text{ m}^2/\text{s}^2 - (15 \text{ m/s})^2}{2(1 \text{ m} - 0 \text{ m})} = -112.5 \text{ m/s}^2$$

$$\Rightarrow F_{net} = F = ma = (60 \text{ kg})(-112.5 \text{ m/s}^2) = -6750 \text{ N}$$

The net force is 6800 N to the left.

(b) Using the same approach as in part **(a)**,

$$F = ma = m\frac{v_f^2 = v_i^2}{2(x_f - x_i)} = (60 \text{ kg})\frac{0 \text{ m}^2/\text{s}^2 - (15 \text{ m/s})^2}{2(0.005 \text{ m})} = -1,350,000 \text{ N}$$

The net force is 1.4×10^6 N to the left.

(c) The passenger's weight is $mg = (60 \text{ kg})(9.80 \text{ m/s}^2) = 590$ N. The force in part **(a)** is 11 times the passenger's weight. The force in part **(b)** is 2300 times the passenger's weight.

Assess: An acceleration of $11g$ is well within the capability of the human body to withstand. A force of 2300 times the passenger's weight, on the other hand, would surely be catastrophic.

P5.59. Prepare: The child is not accelerating in the y-direction, so we can use Equation 5.1 for the forces perpendicular to the incline.
Solve:

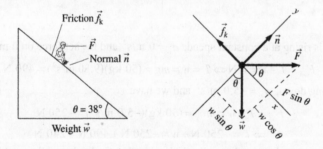

There are three forces with components in the y-direction, the normal force, the weight of the child, and the force of the rope. Equation 5.1 gives

$$n + F\sin(\theta) - w\cos(\theta) = 0$$

Solving for the normal force, we have

$$n = w\cos(\theta) - F\sin(\theta) = mg\cos(\theta) - F\sin(\theta) = (23 \text{ kg})(9.80 \text{ m/s}^2)\cos(38°) - (30 \text{ N})\sin(38°) = 160 \text{ N}$$

Assess: This is less than the child's weight, as expected. Note that the force from the rope acts to decrease the normal force on the child since it tends to pull the child away from the incline.

P5.63. Prepare: We will use Newton's second law with $\mu_k = 0.20$ given in Table 5.2. We also orient our coordinate system so the x-axis runs down the ramp.

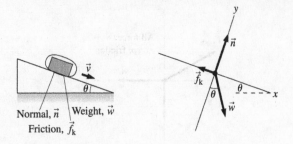

Solve:

$$\Sigma F_x = mg \sin \theta - \mu_k n = 0$$
$$\Sigma F_y = n - mg \cos \theta = 0$$

Solve the second equation for n and insert into the first.

$$mg \sin \theta - \mu_k mg \cos \theta = 0$$
$$\tan \theta = \mu_k \Rightarrow \theta = \tan^{-1}(\mu_k) = \tan^{-1}(0.20) = 11°$$

Assess: The slope of the ramp seems shallow, but probably OK for the low coefficient of friction given.

P5.65. Prepare: The book is in static equilibrium so Equation 5.1 can be applied. The maximum static frictional force the person can exert will determine the heaviest book he can hold.
Solve: Consider the free-body diagram below. The force of the fingers on the book is the reaction force to the normal force of the book on the fingers, so is exactly equal and opposite the normal force on the fingers.

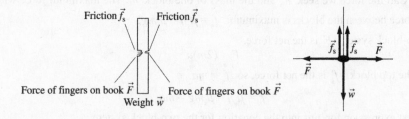

The maximal static friction force will be equal to $f_{s\,max} = \mu_s n = (0.80)(6.0\ \text{N}) = 4.8\ \text{N}$. The frictional force is exerted on both sides of the book. Considering the forces in the y-direction, the weight supported by the maximal frictional force is

$$w = f_{s\,max} + f_{s\,max} = 2 f_{s\,max} = 9.6\ \text{N}$$

We now find the mass of a 9.6 N book.

$$m = \frac{w}{g} = \frac{9.6\ \text{N}}{9.80\ \text{m/s}^2} = 0.98\ \text{kg}$$

Assess: Note that the force on both sides of the book are exactly equal also because the book is in equilibrium.

P5.67. Prepare: We show below the free-body diagram of the 1 kg block. The block is initially at rest, so initially the friction force is static friction. If the 12 N pushing force is too strong, the box will begin to move up the wall. If it is too weak, the box will begin to slide down the wall. And if the pushing force is within the proper range, the box will remain stuck in place.

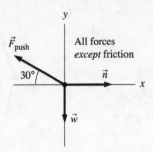

Solve: First, let's evaluate the sum of all the forces *except* friction:

$$\sum F_x = n - F_{push} \cos 30° = 0 \text{ N} \Rightarrow n = F_{push} \cos 30°$$

$$\sum F_y = F_{push} \sin 30° - w = F_{push} \sin 30° - mg = (12 \text{ N}) \sin 30° - (1 \text{ kg})(9.80 \text{ m/s}^2) = -3.8 \text{ N}$$

In the first equation we have utilized the fact that any motion is parallel to the wall, so $a_x = 0 \text{ m/s}^2$.

The two forces in the second y-equation add up to -3.8 N. This means the static friction force will be able to prevent the box from moving if $f_s = +3.8$ N. Using the x-equation we get

$$f_{s\,max} = \mu_s n = \mu_s F_{push} \cos 30° = 5.2 \text{ N}$$

where we used $\mu_s = 0.5$ for wood on wood. The static friction force $\vec{f_s}$ needed to keep the box from moving is *less* than $f_{s\,max}$. Thus the box will stay at rest.

P5.71. Prepare: Call the force we seek F, and the mass of one block m. The maximum force without slippage is when the friction force between the blocks is maximum: $f_s = \mu_s n = \mu_s mg$.

Solve: For the two-block system F is the net force.

$$F = (2m)a$$

Considering only the top block, f_s is the net force, so $f_s = ma$.

$$f_s = \mu_s n = \mu_s mg = ma$$

Now insert our latest expression for ma into the equation for the two-block system.

$$F = 2ma = 2(\mu_s mg) = 2(\mu_s mg) = 2(0.35)(9.80 \text{ m/s}^2)(2.0 \text{ kg}) = 14 \text{ N}$$

Assess: 14 N seems reasonable.

P5.73. Prepare: The Ping-Pong ball when shot straight up is subject to a net force that is the resultant of the weight and drag force vectors, both acting vertically downward. On the other hand, for the ball's motion straight down, the ball is subject to a net force that is the resultant of the weight and drag force vectors, the former in the downward and the latter in the upward direction. An overview of a pictorial representation and a free-body diagram are shown below. The Ping-Pong ball experiences a drag force equal to $\frac{1}{4}\rho A v^2$, as modeled in the text with v_{term} as the terminal velocity.

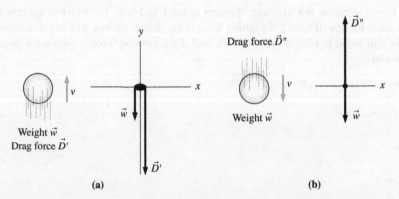

(a) (b)

Solve: (a) Imagine the ball falling at its terminal speed. The ball's weight is directed down and the resistive drag force is directed up. The net force is zero because the magnitude of the drag force is equal to the magnitude of the weight, $D = w$. When the ball is shot upward at twice the terminal speed, the drag force is four times the terminal drag force. That is, $D' = 4D = 4w$.

Since all the forces are down, the y-component of Newton's second law is

$$\sum F_y = -D' - w = -4w - w = -5mg = ma \Rightarrow a = -5g = -49 \text{ m/s}^2$$

(b) The ball is initially shot downward. Therefore D'' is upward but w is down. Again $D'' = 4D$ and the y-component of Newton's second law is

$$\sum F_y = D'' - w = 4w - w = 3mg = ma \Rightarrow a = 3g = 29 \text{ m/s}^2$$

That is, the ball initially decelerates at $3g$ but as v becomes smaller, the drag force approaches the weight so the deceleration goes to zero and v approaches v_{term}.

Assess: D' is very large and with w yields a large initial deceleration when the ball is shot up. When the ball is shot down w opposes D'' so the ball decelerates at a lesser rate.

6

CIRCULAR MOTION, ORBITS, AND GRAVITY

Q6.3. Reason: Because the centripetal acceleration is given by $a = v^2/r$, if the speed is zero then the centripetal acceleration is zero. So the answer is no.
Assess: However, the particle may have a nonzero tangential acceleration at the instant its speed is zero (this would ensure that the particle doesn't stay at rest).

Q6.5. Reason: The discussion in the section on maximum walking speed leads to the equation $v_{max} = \sqrt{gr}$ where r is the length of the leg. For a leg as short as a chickadee's this produces a walking speed that is simply too slow to be practical, so they hop or fly.
Assess: The longer the leg the greater the maximum walking speed, and the formula produces reasonable walking speeds for pheasants.

Q6.9. Reason: (a) The moon's orbit around the earth is fairly circular, and it is the gravitational force of the earth on the moon that provides the centripetal force to keep the moon in its circular motion.
(b) The riders in the Gravitron carnival ride (Section 6.3) have a centripetal acceleration caused by the normal force of the walls on them.
Another example would be the biological sample in a centrifuge. The test tube walls exert a normal force on the sample toward the center of the circle.
Assess: The point is that centripetal forces are not a new *kind* of force; it is just the name we give to the force (or sum of forces) that points toward the center of the circle and keeps the object from flying off in a straight line.

Q6.11. Reason: The car is traveling along a circle and so it must have centripetal acceleration which points downward. From Newton's second law, if an object is accelerating downward, the total force on the object must be downward. The answer is C because only there is the downward force (the weight of the car) greater than the upward force (the normal force on the car) so that the total force is downward.
Assess: It makes sense that the normal force on the car would be less than the weight of the car because, from experience, you know that you feel lighter going over a hill in your car and normal force tells you how heavy you feel. In the same way, the normal force on the car will be less than its weight.

Q6.15. Reason: The radius of the loop decreases as the carts enter and exit the loop. The centripetal acceleration is smaller for larger radius loops and larger for smaller radius loops. This means the centripetal acceleration increases from a minimum at the entry to the loop to a maximum at the top of the loop and then decreases as the cars exit the loop. This prevents a sudden change of acceleration, which can be painful. This also limits the largest accelerations to the top of the loop, so that riders only experience the maximum acceleration for a portion of the trip.
Assess: This is reasonable. If the cars entered a small radius loop directly, the centripetal acceleration would increase suddenly.

Q6.17. Reason: When we walk on the ground we push off with one foot while pivoting on the other; the weight force brings us back down from the push-off for the next step. In an orbiting station, which is in free fall along with the astronaut, after one foot pushes off there isn't a force to bring the astronaut back to the "floor" for the next step; the first push-off sends the astronaut across the cabin.

Assess: If the spacecraft is designed to rotate to provide an artificial gravity then one can walk fairly normally around on the inside; "up" would be toward the center of the circular motion, "down" would be "out"; but that probably isn't the origin of the phrase "down and out."

Q6.21. Reason: Originally, the ball is going around once every second. When the ball is sped up so that it goes around once in only half a second, it is moving twice as fast. Consequently its acceleration, which is given by $a = \omega^2 r$ will be four times as great. From Newton's second law, force is directly proportional to acceleration, so if we multiply the acceleration by 4, we must multiply the tension by 4. Thus the tension in the string will be four times as great, or 24 N. The answer is D.

Assess: This accords with our experience that when we swing an object around a circle, as the speed increases, the tension in the string increases.

Q6.23. Reason: The static friction is directed centripetally and is the net force. The radius of the turn is 95 m.

$$F_{net} = ma = m\frac{v^2}{r} = (610 \text{ kg})\frac{(68 \text{ m/s})^2}{95 \text{ m}} = 30,000 \text{ N}$$

The correct choice is E.

Assess: This large friction force is only possible if the wings help push the car into the track.

Q6.25. Reason: A free-body diagram follows.

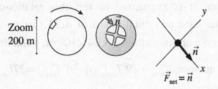

The centripetal force acts toward the center of the circle and is provided entirely by the normal force of the floor of the station.

$$w_{app} = n = m\omega^2 r$$

In order for the occupants to feel as if they are in an environment with an artificial gravity of 1-g, the centripetal acceleration must equal g.

$$\omega^2 r = g$$

Solving for the angular velocity

$$\omega = \sqrt{\frac{g}{r}} = \sqrt{\frac{9.80 \text{ m/s}^2}{100 \text{ m}}} = 0.313 \text{ rad/s}$$

Additional significant figures have been kept in this intermediate result. The period of the rotation is given by

$$T = \frac{2\pi \text{ rad}}{\omega} = \frac{2\pi \text{ rad}}{0.313 \text{ rad/s}} = 20 \text{ s}$$

The correct choice is B.

Assess: Note that quite a short period is required even though the station has a large radius.

Q6.27. Reason: The speed of a satellite in low orbit is $v = \sqrt{gr}$. Use ratios to find $v_{\text{Jup}}/v_{\text{Earth}}$.

$$\frac{v_{\text{Jup}}}{v_{\text{Earth}}} = \frac{\sqrt{g_{\text{Jup}} r_{\text{Jup}}}}{\sqrt{g_{\text{Earth}} r_{\text{Earth}}}} = \frac{\sqrt{(2.5 g_{\text{Earth}})(11 r_{\text{Earth}})}}{\sqrt{g_{\text{Earth}} r_{\text{Earth}}}} = \sqrt{(2.5)(11)} = 5.2$$

The speed of a satellite in low Jupiter orbit is 5.2 times the speed of a satellite in low Earth orbit, so the correct choice is A.

Assess: Both factors made the speed greater around Jupiter.

Q6.31. Reason: We need to use Equation 6.22 (also known as Kepler's Third Law) because it relates the orbital period T to the orbital radius r. We are given that $r_2 = 4r_1$.

Write Equation 6.22 for each planet (write planet 2 first) and then divide the two equations:

$$T_2^2 = \left(\frac{4\pi^2}{GM}\right) r_2^3$$

$$T_1^2 = \left(\frac{4\pi^2}{GM}\right) r_1^3$$

$$\frac{T_2^2}{T_1^2} = \frac{r_2^3}{r_1^3}$$

$$\frac{T_2^2}{T_1^2} = \frac{(4r_1)^3}{r_1^3}$$

Multiply both sides by T_1^2 and cancel r_1^3:

$$T_2^2 = T_1^2 (4)^3$$

Take square roots:

$$T_2 = T_1 \sqrt{(4)^3} = T_1 \sqrt{64} = 8T_1$$

The correct choice is D.

Assess: When the orbital radius quadruples, the period increases by a factor of eight because planet 2 has not only farther to go, but also moves slower. It is instructive to test this relationship with real data. According to Example 6.15, communication satellites have an orbital radius of 4.22×10^7 m and we know from the table inside the back cover of the book that the moon's orbital radius is 3.84×10^8 m. Combining these, we have $r_{moon} \approx 9r_{satellite}$, so using the math above with the new number, $T_{moon} \approx \sqrt{9^3}\, T_{satellite} = \left(\sqrt{9}\right)^3 T_{satellite} = 27 T_{satellite} = 27\,\text{d}$. From Question 6.28, we know that this is the length of one month.

Problems

P6.3. Prepare: Assume uniform circular motion.

Solve: (a) Converting revolutions per minute to revolutions per second

$$\left(33\frac{1}{3}\frac{\text{revolutions}}{\text{minute}}\right)\left(\frac{1 \text{ minute}}{60 \text{ s}}\right) = 0.56 \text{ rev/s}$$

(b) Using the equation from the text

$$T = \frac{1}{f} = \frac{1}{0.56 \text{ rev/s}} = 1.8 \text{ s}$$

Assess: This seems reasonable, if you're old enough to remember LPs. They are making a comeback now.

P6.5. Prepare: We are asked to find period, speed and acceleration. Period and frequency are inverses according to the chapter. To find speed we need to know the distance traveled by the speck in one period. Then the acceleration is given by $a = v^2/r$.

Solve: (a) The disk's frequency can be converted as follows:

$$10,000\frac{\text{rev}}{\text{min}} = 10,000\frac{\text{rev}}{\text{min}}\left(\frac{1 \text{ min}}{60 \text{ sec}}\right) = 167\frac{\text{rev}}{\text{sec}} \approx 170\frac{\text{rev}}{\text{sec}}$$

The period is the inverse of the frequency:

$$T = \frac{1}{f} = \frac{1}{167 \text{ rev/s}} = 6.0 \text{ ms}$$

(b) The speed of the speck equals the circumference of its orbit divided by the period:

$$v = \frac{2\pi r}{T} = \frac{2\pi (6.0 \text{ cm})}{6.00 \text{ ms}} \left(\frac{1000 \text{ ms}}{1 \text{ s}} \right) \left(\frac{1 \text{ m}}{100 \text{ cm}} \right) = 62.8 \text{ m/s},$$

which rounds to 63 m/s.

(c) From Equation 3.23, the acceleration of the speck is given by v^2/r:

$$a = \frac{v^2}{r} = \frac{(62.8 \text{ m/s})^2}{6.0 \text{ cm}} \left(\frac{100 \text{ cm}}{1 \text{ m}} \right) = 65,700 \text{ m/s}^2,$$

which rounds to 66,000 m/s². In units of g, this is as follows:

$$65,700 \text{ m/s}^2 = 65,700 \text{ m/s}^2 \left(\frac{1g}{9.80 \text{ m/s}^2} \right) = 6,700g$$

Assess: The speed and acceleration of the edge of a CD are remarkable. The speed, 63 m/s, is about 140 mi/hr. As you will learn in chapter 4, very large forces are necessary to create large accelerations like 6,700 g.

P6.9. Prepare: The pebble is a particle rotating around the axle in a circular orbit. To convert units from rev/s to rad/s, we note that 1 rev = 2π rad.

Solve: The pebble's angular velocity $\omega = (3.0 \text{ rev/s})(2\pi \text{ rad/rev}) = 18.85 \text{ rad/s}$. The speed of the pebble as it moves around a circle of radius $r = 30 \text{ cm} = 0.30 \text{ m}$ is

$$v = \omega r = (18.85 \text{ rad/s})(0.30 \text{ m}) = 5.65 \text{ m/s} = 5.7 \text{ m/s}$$

The centripetal acceleration is

$$a = \frac{v^2}{r} = \frac{(5.65 \text{ m/s}^2)}{0.30 \text{ m}} = 110 \text{ m/s}^2$$

Assess: These numbers seem reasonable.

P6.11. Prepare: The pilot is assumed to be a particle.

Solve: Since $a_r = v^2/r$, we have

$$v^2 = a_r r = (98 \text{ m/s}^2)(12 \text{ m}) \Rightarrow v = 34 \text{ m/s}$$

Assess: 34 m/s ≈ 76 mph is a large yet understandable speed.

P6.15. Prepare: Treat the block as a particle attached to a massless string that is swinging in a circle on a frictionless table. A pictorial representation of the block, its free-body diagram, and a list of values are shown below. We will use equations from the text and work with SI units.

Solve: (a) The angular velocity and linear speed are

$$\omega = 75 \frac{\text{rev}}{\text{min}} \times \frac{2\pi \text{ rad}}{1 \text{ rev}} = 471.2 \text{ rad/min} \quad v = r\omega = (0.5 \text{ m})(471.2 \text{ rad/min}) \times \frac{1 \text{ min}}{60 \text{ s}} = 3.93 \text{ m/s} \approx 3.9 \text{ m/s}$$

(b) Newton's second law is

$$\Sigma F_r = T = \frac{mv^2}{r}$$

Thus

$$T = (0.200 \text{ kg})\frac{(3.93 \text{ m/s})^2}{0.5 \text{ m}} = 6.2 \text{ N}$$

P6.17. Prepare: We are using the particle model for the car in uniform circular motion on a flat circular track. There must be friction between the tires and the road for the car to move in a circle. A pictorial representation of the car, its free-body diagram, and a list of values are shown below.

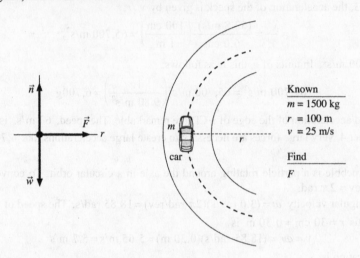

Known
$m = 1500$ kg
$r = 100$ m
$v = 25$ m/s

Find
F

Solve: The equation in the text gives the centripetal acceleration

$$a = \frac{v^2}{r} = \frac{(25 \text{ m/s})^2}{100 \text{ m}} = 6.25 \text{ m/s}^2$$

The acceleration points to the center of the circle, so the net force is

$$\vec{F} = m\vec{a} = (1500 \text{ kg})(6.25 \text{ m/s}^2, \text{ toward center}) = (9400 \text{ N, toward center})$$

This force is provided by static friction:

$$f_s = F_r = 9400 \text{ N}$$

P6.19. Prepare: We can calculate the ball's centripetal acceleration and the centripetal force.
Solve: Refer to the following figure.

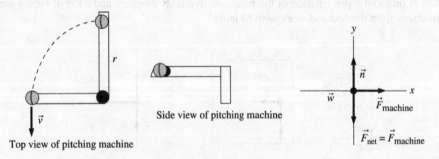

Top view of pitching machine

Side view of pitching machine

(a) Converting the velocity of the ball to meters per second, we have

$$v = (85 \text{ mph})\left(\frac{0.447 \text{ m/s}}{1 \text{ mph}}\right) = 38 \text{ m/s}$$

The centripetal acceleration of the ball is then

$$a = \frac{v^2}{r} = \frac{(38 \text{ m/s}^2)}{0.85 \text{ m}} = 1.7 \times 10^3 \text{ m/s}^2$$

(b) From the free-body diagram in the figure above, the net force on the ball is in the centripetal direction and so is equal to the centripetal force on the ball.

$$F_{net} = ma = (0.144 \text{ kg})(1700 \text{ m/s}^2) = 240 \text{ N}$$

Assess: The centripetal acceleration is large. The centripetal force needed during the launch of the ball is about 54 pounds.

P6.21. Prepare: The force exerted by the wall of the truck on the box provides the centripetal force so we need $F_{wall} = mv^2/r$. The figure shows the box clinging to the left wall of the truck bed.

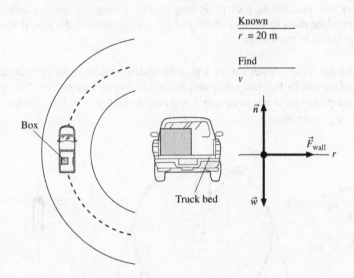

Known
r = 20 m

Find
v

Solve: Since the force exerted by the wall equals the weight of the box, we can write: $F_{wall} = mv^2/r = mg$. If we solve this equation for v, we get:

$$v = \sqrt{gr} = \sqrt{(9.80 \text{ m/s}^2)(20 \text{ m})} = 14 \text{ m/s}$$

The truck needs to travel at 14 m/s.

Assess: This is reasonable because we know from experience that at typical vehicular speeds, the forces on our bodies (exerted by the seat belt) can be large compared to our weight, especially for sharp turns.

P6.25. Prepare: We will calculate the critical speed of the rock in the bucket.
Solve: A free-body diagram is shown.

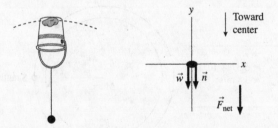

At the top of the circle, the only forces on the rock are the weight of the rock and the normal force of the bottom of the bucket on the rock. Both these forces are directed toward the center of the circle. Newton's second law gives

$$F_{net} = n + w = \frac{mv^2}{r}$$

Solving for the normal force,

$$n = \frac{mv^2}{r} - mg$$

The normal force is equal to zero when the velocity has a magnitude equal to the critical speed. Solving for v when $n = 0$ N in the equation above,

$$v_c = \sqrt{rg} = \sqrt{(1.1 \text{ m})(9.80 \text{ m/s}^2)} = 3.3 \text{ m/s}$$

If the magnitude of the velocity of the rock is just equal to the critical speed, the normal force is exactly zero and the rock is on the verge of leaving the bottom of the bucket.

Assess: It doesn't matter how massive the rock is, as long as it is moving at a speed greater than the critical speed. The result is independent of the mass of the rock. Note that the critical speed is the lowest speed that the rock can be traveling to remain in contact with the bucket.

P6.27. Prepare: Model the roller coaster car as a particle undergoing uniform circular motion along a loop. A pictorial representation of the car, its free-body diagram, and a list of values are shown. Note that the normal force $\vec{n}$ of the seat pushing on the passenger is the passenger's apparent weight, and in this problem the apparent weight is equal to the true weight: $w_{app} = n = mg$.

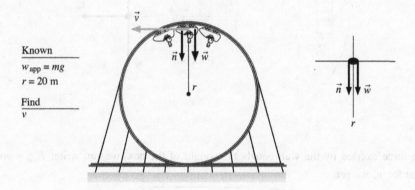

Solve: We have

$$\Sigma F = n + w = \frac{mv^2}{r} = mg + mg \Rightarrow v = \sqrt{2rg} = \sqrt{2(20 \text{ m})(9.80 \text{ m/s}^2)} = 20 \text{ m/s}$$

Assess: A speed of 20 m/s or 44 mph on a roller coaster ride is reasonable. The mass cancels out of the calculation.

P6.31. Prepare: Assume the radius of the satellite's orbit is about the same as the radius of Mars itself.

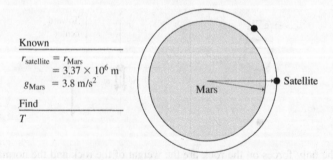

As a preliminary calculation, compute the angular velocity of the satellite:

$$\omega = \frac{2\pi}{T} = \frac{2\pi \text{ rad}}{110 \text{ min}} \left(\frac{1 \text{ min}}{60 \text{ s}} \right) = 9.52 \times 10^{-4} \text{ rad/s}$$

Solve: Since $T = \dfrac{2\pi}{\omega}$ and $a = \omega^2 r \Rightarrow \omega = \sqrt{\dfrac{a}{r}}$, then

$$T = \frac{2\pi}{\omega} = \frac{2\pi}{\sqrt{\dfrac{a}{r}}} = \frac{2\pi}{\sqrt{\dfrac{3.8 \text{ m/s}^2}{3.37 \times 10^6 \text{ m}}}} = 5900 \text{ s}$$

This answer is equal to about 99 min.

Assess: This is between the orbital period for a satellite in low earth orbit and one in low moon orbit, which sounds right.

P6.33. Prepare: Call the mass of the star M. Write Newton's law of gravitation for each planet.

$$F_1 = \frac{GMm_1}{r_1^2}$$

$$F_2 = \frac{GMm_2}{r_2^2} = \frac{GM(2m_1)}{(2r_1)^2}$$

Solve: Divide the two equations to get the ratio desired.

$$\frac{F_2}{F_1} = \frac{\frac{GM(2m_1)}{(2r_1)^2}}{\frac{GMm_1}{r_1^2}} = \frac{1}{2}$$

Assess: The answer is expected. Even with twice the mass, because the radius in the denominator is squared, we expect the force on planet 2 to be less than the force on planet 1.

P6.39. Prepare: Model the sun (s), the earth (e), and the moon (m) as spherical masses.

Solve: (a) $F_{\text{s on e}} = \dfrac{Gm_s m_e}{r_{s-e}^2} = \dfrac{(6.67 \times 10^{-11} \text{ N} \cdot \text{m}^2/\text{kg}^2)(1.99 \times 10^{30} \text{ kg})(5.98 \times 10^{24} \text{ kg})}{(1.50 \times 10^{11} \text{ m})^2} = 3.53 \times 10^{22} \text{ N}$

(b) $F_{\text{m on e}} = \dfrac{GM_m M_e}{r_{m-e}^2} = \dfrac{(6.67 \times 10^{-11} \text{ N} \cdot \text{m}^2/\text{kg}^2)(7.36 \times 10^{22} \text{ kg})(5.98 \times 10^{24} \text{ kg})}{(3.84 \times 10^8 \text{ m})^2} = 1.99 \times 10^{20} \text{ N}$

(c) The moon's force on the earth as a percent of the sun's force on the earth is

$$\left(\frac{1.99 \times 10^{20} \text{ N}}{3.53 \times 10^{22} \text{ N}} \right) \times 100 = 0.564\%$$

P6.41. Prepare: Model Mars (m) and Jupiter (J) as spherical masses.

Solve: (a) $g_{\text{Mars surface}} = \dfrac{(6.67 \times 10^{-11} \text{ N} \cdot \text{m}^2/\text{kg}^2)(6.42 \times 10^{23} \text{ kg})}{(3.37 \times 10^6 \text{ m})^2} = 3.77 \text{ m/s}^2$

(b) $g_{\text{Jupiter surface}} = \dfrac{GM_J}{R_J^2} = \dfrac{(6.67 \times 10^{-11} \text{ N} \cdot \text{m}^2/\text{kg}^2)(1.90 \times 10^{27} \text{ kg})}{(6.99 \times 10^7 \text{ m})^2} = 25.9 \text{ m/s}^2$

P6.43. Prepare: We can use the equation for the speed of a satellite in a circular orbit. Assume the two satellites are spherical masses with center-to-center separations of R_A and R_B from a planet such that $r_B = 2r_A$ and $m_B = 2m_A$.

Solve: (a) From Equation 6.21, $v_A = \sqrt{GM/r_A}$ and $v_B = \sqrt{GM/r_B}$, where M is the planet's mass. So,

$$\frac{v_B}{v_A} = \sqrt{\frac{r_A}{r_B}} \Rightarrow v_B = 10{,}000 \text{ m/s} \sqrt{\frac{1}{2}} = 7000 \text{ m/s}$$

Assess: Note that the mass of a planet does not figure in the equation.

P6.45. Prepare: Model the sun (s) as a spherical mass and the asteroid (a) as a point particle. The asteroid, having mass m_a and velocity v_a, orbits the sun in a circle of radius r_a. The asteroid's time period is $T_a = 5.0$ earth years $= 1.5779 \times 10^8$ s.

Solve: The gravitational force between the sun (mass $= M_s$) and the asteroid provides the centripetal acceleration required for circular motion.

$$\frac{GM_s m_a}{r_a^2} = \frac{m_a v_a^2}{r_a} \Rightarrow \frac{GM_s}{r_a} = \left(\frac{2\pi r_a}{T_a}\right)^2 \Rightarrow r_a = \left(\frac{GM_s T_a^2}{4\pi^2}\right)^{1/3}$$

Substituting $G = 6.67 \times 10^{-11}$ N·m^2/kg^2, $M_s = 1.99 \times 10^{30}$ kg, and the time period of the asteroid, we obtain $r_a = 4.37 \times 10^{11}$ m. The velocity of the asteroid in its orbit will therefore be

$$v_a = \frac{2\pi r_a}{T_a} = \frac{(2\pi)(4.37 \times 10^{11} \text{ m})}{1.5779 \times 10^8 \text{ s}} = 1.7 \times 10^4 \text{ m/s}$$

P6.49 Prepare: From the equation for circular orbits we solve for T. Preliminary calculations give 0.0058 au $= 8.70 \times 10^8$ m and $0.13 M_{sun} = 2.59 \times 10^{29}$ kg.

Solve: The speed is

$$T = 2\pi \sqrt{\frac{r^3}{GM}} = 2\pi \sqrt{\frac{(8.70 \times 10^8 \text{ m})^3}{(6.67 \times 10^{-11} \text{ N·m}^2/\text{kg}^2)(2.59 \times 10^{29} \text{ kg})}} = 11 \text{ h}$$

Assess: This is an extremely short year. This problem can also be solved using ratios.

P6.53. Prepare: Treat the man as a particle. When at the equator the man undergoes uniform circular motion as the earth rotates.

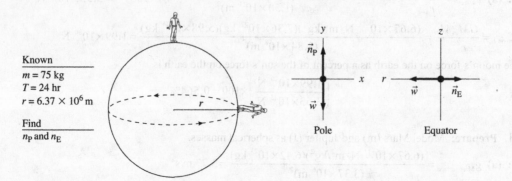

Known
$m = 75$ kg
$T = 24$ hr
$r = 6.37 \times 10^6$ m

Find
n_P and n_E

Pole   Equator

Solve: The scale reads the man's apparent weight $w_{app} = n$, the force of the scale pushing up against his feet. At the North Pole, where the man is in static equilibrium,

$$n_P = (w_{app})_P = mg = 735 \text{ N}$$

At the equator, there must be a net force toward the center of the earth to keep the man moving in a circle. In the radial direction

$$\Sigma F = w - n_E = m\omega^2 r \Rightarrow n_E = (w_{app})_E = mg - m\omega^2 r = (w_{app})_P - m\omega^2 r$$

So the equator scale reads less than the North Pole scale by the amount $m\omega^2 r$.

The angular velocity of the earth is

$$\omega = \frac{2\pi}{T} = \frac{2\pi \text{ rad}}{24 \text{ h} \times (3600 \text{ s}/1 \text{ h})} = 7.27 \times 10^{-5} \text{ rad/s}$$

Thus the North Pole scale reads more than the equator scale by

$$m\omega^2 r = (75 \text{ kg})(7.27 \times 10^{-5} \text{ rad/s})^2 (6.37 \times 10^6 \text{ m}) = 2.5 \text{ N}$$

Assess: The man at the equator appears to have lost approximately 0.25 kg or about 1/2 lb.

P6.55. Prepare: Model the ball as a particle which is in a vertical circular motion. A visual overview of the ball's vertical motion is shown in the following pictorial representation, free-body diagram, and list of values. The tension in the string causes the centripetal acceleration needed for the ball's circular motion.

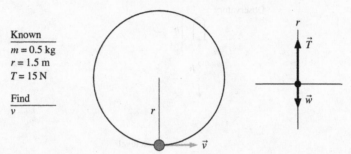

Known
$m = 0.5$ kg
$r = 1.5$ m
$T = 15$ N

Find
v

Solve: At the bottom of the circle,

$$\sum F_{\text{bottom}} = T - w = \frac{mv^2}{r} \Rightarrow (15\ \text{N}) - (0.5\ \text{kg})(9.8\ \text{m/s}^2) = \frac{(0.5\ \text{kg})v^2}{(1.5\ \text{m})} \Rightarrow v = 5.5\ \text{m/s}$$

Assess: A speed of 5.5 m/s or 12 mph is reasonable for the ball attached to a string.

P6.59. Prepare: Since the hanging block is at rest, the total force on it is zero. The two forces are the tension in the string, T, and the weight of the puck, $-mg$. Since the revolving puck is moving at constant speed in a circle, the total force on the puck is the centripetal force. We must write the equations and solve them.
Solve: The total force on the block is $T - mg$. From Newton's second law, the total force is zero so we write:

$$T = mg = (1.20\ \text{kg})(9.80\ \text{m/s}^2) = 11.8\ \text{N}$$

The centripetal acceleration of the puck is caused by the tension in the string, so $mv^2 / r = T$. We solve this to obtain:

$$v = \sqrt{Tr / m} = \sqrt{(11.8\ \text{N})(0.50\ \text{m})/(0.20\ \text{kg})} = 5.4\ \text{m/s}$$

The puck must rotate at a speed of 5.4 m/s.
Assess: It is remarkable that a block can be supported by a puck moving horizontally. But both the puck and the block are able to pull on the string—the block pulls downward on one end and the puck pulls outward on the other end. The relatively small mass of the puck is compensated by its high speed of 5.4 m/s.

P6.61. Prepare: Treat the car as a particle which is undergoing circular motion. The car is in circular motion with the center of the circle below the car. A visual overview of the car's circular motion is shown below in the following pictorial representation, free-body diagram, and list of values.

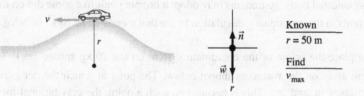

Known
$r = 50$ m

Find
v_{max}

Solve: Newton's second law at the top of the hill is

$$F_{\text{net}} = \sum F_y = w - n = mg - n = ma = \frac{mv^2}{r} \Rightarrow v^2 = r\left(g - \frac{n}{m}\right)$$

This result shows that maximum speed is reached when $n = 0$ and the car is beginning to lose contact with the road. Then,

$$v_{\text{max}} = \sqrt{rg} = \sqrt{(50\ \text{m})(9.80\ \text{m/s}^2)} = 22\ \text{m/s}$$

Assess: A speed of 22 m/s is equivalent to 50 mph, which seems like a reasonable value.

P6.65. Prepare: Model the earth (e) as a spherical mass. We will take the free-fall acceleration to be 9.83 m/s^2 and $R_e = 6.37 \times 10^6$ m. A pictorial representation of the situation is shown.

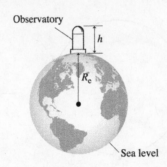

Observatory

h

R_e

Sea level

Solve: $g_{observatory} = \dfrac{GM_e}{(R_e + h)^2} = \dfrac{GM_e}{R_e^2(1 + \frac{h}{R_e})^2} = \dfrac{g_{earth}}{(1 + \frac{h}{R_e})^2} = (9.83 - 0.0075) \text{ m/s}^2$

Here $g_{earth} = GM_e/R_e^2$ is the free-fall acceleration. Solving for h,

$$h = \left(\sqrt{\frac{9.83}{9.8225}} - 1 \right) R_e = 2400 \text{ m}$$

Assess: This altitude is relative to the sea level and is at reasonable altitude.

P6.69. Prepare: According to the discussion in Section 6.2, the maximum walking speed is $v_{max} = \sqrt{gr}$. The astronaut's leg is about 0.70 m long whether on earth or on Mars, but g will be difficult. Use the equation to find g_{Mars}.

We look up the required data in the astronomical table: $m_{Mars} = 6.42 \times 10^{23}$ kg, and $R_{Mars} = 3.37 \times 10^6$ m. In part **(b)** we'll make the same assumption as in the text: The length of the leg $r = 0.70$ m.

Solve: (a)

$$g_{Mars} = \frac{GM_{Mars}}{(R_{Mars})2} = \frac{(6.67 \times 10^{-11} \text{ N} \cdot \text{m}^2/\text{kg}^2)(6.42 \times 10^{23} \text{ kg})}{(3.37 \times 10^6 \text{ m})^2} = 3.77 \text{ m/s}^2 \approx 3.8 \text{ m/s}^2$$

(b)

$$v_{max} = \sqrt{gr} = \sqrt{(3.77 \text{ m/s}^2)(0.70 \text{ m})} = 1.6 \text{ m/s}$$

Assess: The answer is about 3.6 mph, or about 60% of the speed the astronaut could walk on the earth. This is reasonable on a smaller celestial body. Astronauts may adopt a hopping gait like some did on the moon.

Carefully analyze the units in the preliminary calculation to see that g ends up in m/s^2 or N/kg.

P6.71. Prepare: We place the origin of the coordinate system on the 20 kg sphere (m_1). The sphere (m_2) with a mass of 10 kg is 20 cm away on the x-axis, as shown below. The point at which the net gravitational force is zero must lie between the masses m_1 and m_2. This is because on such a point, the gravitational forces due to m_1 and m_2 are in opposite directions. As the gravitational force is directly proportional to the two masses and inversely proportional to the square of distance between them, the mass m must be closer to the 10-kg mass. The small mass m, if placed either to the left of m_1 or to the right of m_2, will experience gravitational forces from m_1 and m_2 pointing in the same direction, thus always leading to a nonzero force.

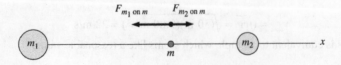

$F_{m_1 \text{ on } m}$ $F_{m_2 \text{ on } m}$

m_1 m m_2 x

Solve:

$$F_{m_1\,on\,m} = F_{m_2\,on\,m} \Rightarrow G\frac{m_1 m}{x^2} = G\frac{m_2 m}{(0.20-x)^2} \Rightarrow \frac{20}{x^2} = \frac{10}{(0.20-x)^2} \Rightarrow 10x^2 - 8x + 0.8 = 0$$

The value $x = 68.3$ cm is unphysical in the current situation, since this point is not between m_1 and m_2. Thus, the point $(x, y) = (11.7\text{ cm}, 0\text{ cm}) \approx (12\text{ cm}, 0\text{ cm})$ is where a small mass is to be placed for a zero gravitational force.

P6.73. Prepare: Model Mars (m) and Phobos as spherical masses.
Solve: The period of a satellite orbiting a planet of mass M_m is

$$T^2 = \left(\frac{4\pi^2}{GM_m}\right) r^3$$

Thus we can use Phobos's orbit to find the mass of Mars:

$$M_m = \frac{4\pi^2 r^3}{GT^2} = \frac{4\pi^2 (9.4\times10^6\text{ m})^3}{(6.67\times10^{-11}\text{ N}\cdot\text{m}^2/\text{kg}^2)(2.7540\times10^4\text{ s})^2} = 6.5\times10^{23}\text{ kg}$$

Assess: The mass of Mars is 6.42×10^{23} kg. The slight difference is likely due to Phobos's orbit being somewhat noncircular.

P6.75. Prepare: According to the discussion in Section 6.2 the maximum walking speed is $v_{max} = \sqrt{gr}$. The astronaut's leg is about 0.70 m long whether on earth or on Europa, but g will be different.

$$g_{Europa} = \frac{GM_{Europa}}{(R_{Europa})^2} = \frac{(6.67\times10^{-11}\text{ N}\cdot\text{m}^2/\text{kg}^2)(4.8\times10^{22}\text{ kg})}{(3.1\times10^6\text{ m})^2} = 0.333\text{ m/s}^2$$

Solve:

$$v_{max} = \sqrt{gr} = \sqrt{(0.333\text{ m/s}^2)(0.70\text{ m})} = 0.48\text{ m/s}$$

Assess: The answer is about 1 mph or about 1/6 of the speed the astronaut could walk on the earth. This is reasonable on a small celestial body. Astronauts may adopt a hopping gait like some did on the moon.
Carefully analyze the units in the preliminary calculation to see that g ends up in m/s^2 or N/kg.

Q7.3. Reason: By convention, clockwise rotations are negative and counterclockwise rotations are positive. As a result, an angular acceleration that decreases/increases a negative angular velocity is positive/negative. In like manner, an angular acceleration that decreases/increases a positive angular velocity is negative/positive. Knowing this we can establish the situation for each figure. Figure (a) the pulley is rotating clockwise ($\omega = -$), however since the large mass is on the left it is decelerating ($\alpha = +$).

Figure (b) the pulley is rotating counterclockwise ($\omega = +$) and since the large mass is on the left it is accelerating ($\alpha = +$).

Figure (c) the pulley is rotating clockwise ($\omega = -$) and since the large mass is on the right it is accelerating ($\alpha = -$).

Figure (d) the pulley is rotating counterclockwise ($\omega = +$), however since the large mass is on the right it is decelerating ($\alpha = -$).

Assess: It is important to know the sign convention for all physical quantities that are vectors. This is especially important when working with rotational motion.

Q7.5. Reason: The question properly identified where the torques are computed about (the hinge). Torques that tend to make the door rotate counterclockwise in the diagram are positive by convention (general agreement) and torques that tend to make the door rotate clockwise are negative.

(a)   +
(b)   −
(c)   +
(d)   −
(e)   0

Assess: Looking at the diagram we see that $\vec{F}_a$ and $\vec{F}_c$ are parallel and are both creating a negative or counterclockwise torque. But since $\vec{F}_c$ is farther from the hinge, its torque will be greater. A similar argument can be made for $\vec{F}_b$ and $\vec{F}_d$. Note that $\vec{F}_e$ causes no torque since it has no moment arm.

Q7.9. Reason: The torque the student exerts turns the ball and rod in a clockwise direction as seen from the top. Clockwise torques are negative. The student exerts a negative torque.

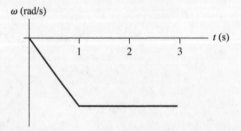

Assess: Note that gravity *does* produce a torque on the ball tending to rotate the ball in the vertical direction, though this torque is balanced by the vertical torque of the rod and pivot keeping the ball moving in a horizontal plane.

Q7.11. Reason: As suggested by the figure, we will assume that the larger sphere is more massive. Then the center of gravity would be at point 1, because if we suspend the dumbbell from point 1 the counterclockwise torque due to the large sphere (large weight times small lever arm) will be equal to the clockwise torque due to the small sphere (small weight times large lever arm).
Assess: Look at the figure and mentally balance the dumbbell on your finger; your finger would have to be at point 1.
The sun-earth system is similar to this except that the sun's mass is so much greater than the earth's that the center of mass (called the barycenter for astronomical objects orbiting each other) is only 450 km from the center of the sun.

Q7.13. Reason: Spin them. Because they would have different moments of inertia ($2/5 MR^2$ for the solid sphere and $2/3 MR^2$ for a thin-walled hollow sphere, with something in between if the wall of the hollow sphere is not particularly thin) their angular accelerations would be different with the same torque acting on them (remember the rotational version of Newton's second law: $\alpha = \tau_{net}/I$). The solid one (with the smaller moment of inertia) would accelerate quicker given the same torque.
Assess: The two balls would have the same linear inertia (having the same mass) so dropping them in free fall, for example, and observing their linear accelerations would not distinguish between them; but the rotational inertia (moment of inertia) is different, and so exerting a net torque on them and measuring the resultant angular acceleration can distinguish between them.

Q7.15 Reason: A review of Section 7.7 in the text will prepare you to answer this question. Let's separate the rolling motion into translation and rotation. Figure (a) below shows the translational velocity of the points of interest. Figure (b) below shows the velocity due to rotation at all points of interest. Finally, Figure (c) below combines the velocity vectors due to both translation and rotation.

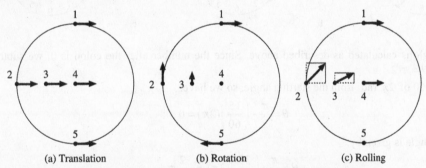

| (a) Translation | (b) Rotation | (c) Rolling |

Looking at the magnitude of the resulting velocity vectors we can write the following:
$$v_1 > v_2 > v_3 > v_4 > v_5 = 0$$
Assess: We know that the magnitudes of $\vec{v}_2$ and $\vec{v}_3$ are greater than the magnitude of $\vec{v}_4$ because their horizontal component is equal to the magnitude of $\vec{v}_4$.

Q7.19. Reason: Since the center of gravity of piece 2 is to the right of the center of gravity of piece 1, the horizontal position of the center of gravity of the two pieces should be between the center of gravity of the two pieces. The same argument applies to the vertical position of the center of gravity of the pieces. The only point that is located between the two centers of gravity is point D.
Assess: Our solution to the problem is based on the fact that we can replace piece 2 with a single mass point (with the same mass as piece 2) located at the center of gravity of piece 2 and we can replace piece 1 with a single mass point (with the same mass as piece 1) located at the center of gravity of piece 1. We now need to find the center of gravity of these two mass points and our knowledge of the physics involved makes us aware that it must be somewhere on the line connecting them. Only point D satisfies this condition.

Q7.21. Reason: Since the diameter of the disk increases by a factor of two, the radius of the disk increases by a factor of two. Assume that the density of the material used is the same and that the hole in the center of the disk is negligible. The total mass of the disk goes up by a factor of four since the volume of the disk increases by a factor of four. The moment of inertia of a disk is given in Table 7.1 and is proportional to the mass of the disk and the square of the radius of the disk. Since the mass goes up by a factor of four and the radius goes up by a factor of two, the moment of inertia of the disk goes up by a total factor of sixteen. The correct choice is D.

Assess: Note that the increase of the total mass of the disk needed to be included in the calculation.

Q7.25. Reason: The skater turns one-and-a-half revolutions in 0.5 s. One-and-a-half revolutions is 3π radians. Her angular velocity is

$$\omega = \frac{\Delta\theta}{\Delta t} = \frac{3\pi \text{ rad}}{0.5 \text{ s}} = 20 \text{ rad/s}$$

The correct choice is D.

Assess: This result is reasonable. She makes three revolutions in one second, which is 6π radians per second.

Problems

P7.1. Prepare: The position of the minute hand is determined by the number after the colon. There are 60 minutes in an hour so the number of minutes after the hour, when divided by 60, gives the fraction of a circle which has been covered by the minute hand. Also, the minute hand starts at $\pi/2$ rad and travels clockwise, thus decreasing the angle. If we get a negative angle, we can make it positive by adding 2π rad.

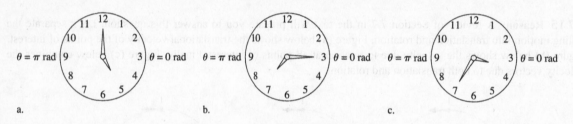

a.   b.   c.

Solve: (a) The angle is calculated as described above. Since the number after the colon is 0, we subtract nothing from $\pi/2$ rad, so $\theta = \pi/2$.

(b) We subtract 15/60 of 2π rad from the starting angle, so we have:

$$\theta = \frac{\pi}{2} - \left(\frac{15}{60}\right)(2\pi) = 0$$

(c) As before, the angle is given by:

$$\theta = \frac{\pi}{2} - \left(\frac{35}{60}\right)(2\pi) = -\frac{2}{3}\pi$$

Since this angle is negative, we can add 2π rad to obtain: $\theta = -2\pi/3 + 2\pi = 4\pi/3$.

Assess: The first two parts make sense from our experience with clocks. In part (a), the minute hand is straight up. In part (b), it points to the right.

P7.5. Prepare: The airplane is to be treated as a particle in uniform circular motion on the equator around the center of the earth. We show the following pictorial representation of the problem and a list of values. To convert radians into degrees, we note that 2π rad $= 360°$.

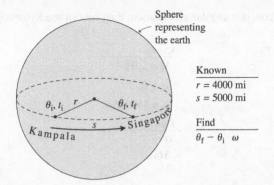

Solve: **(a)** The angle you turn through is

$$\theta_f - \theta_i = \frac{s}{r} = \frac{5000 \text{ mi}}{4000 \text{ mi}} = 1.25 \text{ rad} = 1.25 \text{ rad} \times \frac{180°}{\pi \text{ rad}} = 71.6°$$

So, the angle is 1.3 rad or 72°.

(b) The plane's angular speed is

$$\omega = \frac{\theta_f - \theta_i}{t_f - t_i} = \frac{1.25 \text{ rad}}{9 \text{ h}} = 0.139 \text{ rad/h} = 0.139 \frac{\text{rad}}{\text{h}} \times \frac{1 \text{ h}}{3600 \text{ s}} = 3.9 \times 10^{-5} \text{ rad/s}$$

Assess: An angular displacement of approximately one-fifth of a complete rotation is reasonable because the separation between Kampala and Singapore is approximately one-fifth of the earth's circumference.

P7.7. Prepare: We'll use the equation in the text to compute the angular displacement. We are given $\theta_i = 0.45$ rad and that $\Delta t = 8.0 \text{ s} - 0 \text{ s} = 8.0 \text{ s}$.

We'll do a preliminary calculation to convert $\omega = 78$ rpm into rad/s:

$$78 \text{ rpm} = 78 \frac{\text{rev}}{\text{min}} \left(\frac{2\pi \text{ rad}}{1 \text{ rev}} \right) \left(\frac{1 \text{ min}}{60 \text{ s}} \right) = 8.17 \text{ rad/s}$$

Solve: Solve the equation for θ_f:

$$\theta_f = \theta_i + \omega \Delta t = 0.45 \text{ rad} + (8.17 \text{ rad/s})(8.0 \text{ s}) = 65.8 \text{ rad} = 10.474 \times 2\pi \text{ rad}$$
$$= 10 \times 2\pi \text{ rad} + 0.474 \times 2\pi \text{ rad} = 10 \times 2\pi \text{ rad} + 2.98 \text{ rad}$$

So the speck completed almost ten and a half revolutions. An observer would say the angular position is 3.0 rad (to two significant figures) at $t = 8.0$ s.

Assess: Ask your grandparents if they remember the old records that turned at 78 rpm. They turned quite fast and so the music didn't last long before it was time to turn the record over.

Singles came on smaller records that turned at 45 rpm, and later "long play" (LP) records turned at 33 rpm.

CDs don't have a constant angular velocity, instead they are designed to have constant linear velocity, so the motor has to change speeds. For the old vinyl records the recording had to take into account the changing linear velocity because they had constant angular velocity.

P7.11. Prepare: The smooth rotation assures constant angular velocity.

Solve: **(a)** The second hand has a rotational speed of one revolution per minute, or 1 rpm. We simply need to convert this to SI units.

$$1 \frac{\text{rev}}{\text{min}} = \left(1 \frac{\text{rev}}{\text{min}} \right) \left(\frac{2\pi \text{ rad}}{1 \text{ rev}} \right) \left(\frac{1 \text{ min}}{60 \text{ s}} \right) = 0.105 \frac{\text{rad}}{\text{s}}$$

(b) The tip of the hand has speed

$$v = \omega R = (0.105 \text{ rad/s})(0.0100 \text{ m}) = 0.00105 \text{ m/s}$$

Assess: This is about a millimeter per second, which is about right.

P7.15. Prepare: We assume constant angular acceleration; then we can use Synthesis 7.1.

$$\begin{array}{c} \underline{\text{Known}} \\ \Delta t = 2.0 \text{ s} \\ \omega_0 = 0 \\ \omega_f = 3000 \text{ rpm} \\ \hline \text{Find} \\ \hline \alpha \\ \Delta \theta \end{array}$$

Convert $\Delta \omega$ to rad/s.

$$\Delta \omega = \omega_f - \omega_0 = 3000 \frac{\text{rev}}{\text{min}} \left(\frac{2\pi \text{ rad}}{1 \text{ rev}} \right) \left(\frac{1 \text{ min}}{60 \text{ s}} \right) = 314 \text{ rad/s}$$

Solve: (a)

$$\alpha = \frac{\Delta \omega}{\Delta t} = \frac{314 \text{ rad/s}}{2.0 \text{ s}} \approx 160 \text{ rad/s}^2$$

(b) We'll use Synthesis 7.1.2, using $\omega_0 = 0$.

$$\Delta \theta = \omega_0 \Delta t + \frac{1}{2} \alpha (\Delta t)^2 = \frac{1}{2} (157 \text{ rad/s}^2)(2.0 \text{ s})^2 = 314 \text{ rad}$$

Finally, convert to revolutions:

$$314 \text{ rad} = 314 \text{ rad} \left(\frac{1 \text{ rev}}{2\pi \text{ rad}} \right) = 50 \text{ rev}$$

Assess: 50 rev seems like a lot in 2 s, but it is reasonable with the large angular acceleration and the final angular velocity of 3000 rpm.

P7.17. Prepare: The magnitude of the torque in each case is $\tau = rF$ because $\sin \phi = 1$.

Solve: $\tau_1 = rF$ $\tau_2 = r2F$ $\tau_3 = 2rF$ $\tau_4 = 2r2F$

Examining the above we see that $\tau_1 < \tau_2 = \tau_3 < \tau_4$. Since for each case $\tau = rF$ (because $\sin \phi = 1$), in order to determine the torque we have just kept track of each force (F), the magnitude of the position vector (r) which locates the point of application of the force, and finally the product (rF).

Assess: As expected, both the force and the lever arm contribute to the torque. Larger forces and larger lever arms make larger torques. Case 4 has both the largest force and the largest lever arm, hence the largest torque.

P7.19. Prepare: Torque by a force is defined as $\tau = Fr \sin \phi$ [Equation 7.12], where ϕ is measured counterclockwise from the $\vec{r}$ vector to the $\vec{F}$ vector. The radial line passing through the axis of rotation is shown below by the broken line. We see that the 20 N force makes an angle of $+90°$ relative to the radius vector r_2, but the 30 N force makes an angle of $-90°$ relative to r_1.

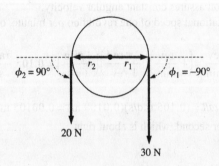

Solve: The net torque on the pulley about the axle is the torque due to the 30 N force plus the torque due to the 20 N force:

$$(30 \text{ N})r_1 \sin \phi_1 + (20 \text{ N})r_2 \sin \phi_2 = (30 \text{ N})(0.02 \text{ m})\sin(-90°) + (20 \text{ N})(0.02 \text{ m})\sin(90°)$$
$$= (-0.60 \text{ N} \cdot \text{m}) + (0.40 \text{ N} \cdot \text{m}) = -0.20 \text{ N} \cdot \text{m}$$

Assess: A negative torque will cause a clockwise rotation of the pulley.

P7.23. Prepare: For both Tom and Jerry $r = 1.5$ m (or half the diameter).
Compute the magnitude of each torque.

$$\tau_{\text{Tom}} = (1.50 \text{ m})(50.0 \text{ N}) \sin 60° = 64.95 \text{ N} \cdot \text{m}$$
$$\tau_{\text{Jerry}} = (1.50 \text{ m})(35.0 \text{ N}) \sin 80° = 51.70 \text{ N} \cdot \text{m}$$

We keep extra significant figures in the intermediate calculations because we will be subtracting two nearly equal numbers.

Solve: (a) The torque due to Tom will be positive because it will tend to produce counterclockwise rotation, while the torque due to Jerry will be negative because it will tend to produce clockwise rotation.

$$\tau = \tau_{\text{Tom}} + \tau_{\text{Jerry}} = 64.95 \text{ N} \cdot \text{m} + (-51.70 \text{ N} \cdot \text{m}) = 13.2 \text{ N} \cdot \text{m}$$

(b) With both Tom's and Jerry's torques tending to produce counterclockwise rotation, they are both positive and so the net torque would be their sum:

$$\tau = \tau_{\text{Tom}} + \tau_{\text{Jerry}} = 64.95 \text{ N} \cdot \text{m} + 51.70 \text{ N} \cdot \text{m} = 116.6 \text{ N} \cdot \text{m}$$

This rounds to $117 \text{ N} \cdot \text{m}$ to 3 significant figures.

Assess: As the difference between the two parts of the problem demonstrates, the direction of the forces really matters. But we can produce torques of the same magnitude with forces of different magnitude by adjusting the angles—even with the same r.

P7.25. Prepare: Knowing that torque may be determined by $\tau = rF_\perp$, that counterclockwise torque is positive and clockwise torque negative we can determine the net torque acting on the bar.

Solve:

$$\tau_{\text{clockwise}} = -(0.25 \text{ m})(8.0 \text{ N}) = -2.0 \text{ N} \cdot \text{m}$$
$$\tau_{\text{counterclockwise}} = (0.75 \text{ m})(10 \text{ N}) = 7.5 \text{ N} \cdot \text{m}$$
$$\tau_{\text{net}} = \tau_{\text{counterclockwise}} + \tau_{\text{clockwise}} = 7.5 \text{ N} \cdot \text{m} - 2.0 \text{ N} \cdot \text{m} = 5.5 \text{ N} \cdot \text{m}$$

Since the net torque is $+5.5 \text{ N} \cdot \text{m}$, the bar will rotate in the counterclockwise direction around the dot.

Assess: The counterclockwise torque had both the larger r and the larger F, so the net torque was also counterclockwise. The numbers also seem reasonable, and the units work out.

P7.29. Prepare: How will you estimate the mass of your arm? Of course different people's arms have different masses, but even different methods of estimating the mass of your specific arm will produce slightly different results. But it is a good exercise, even if we have only one significant figure of precision. One way would be to make some rough measurements. Because the density of your arm is about the density of water, you could fill a garbage can to the brim with water, insert your arm, and weigh the water that overflows. Or you could look at Figure 7.31, which indicates that a guess of $m = 4.0$ kg for a person whose total mass is 80 kg is good.
The gravitational force on a 4.0 kg object is $w = mg(4.0 \text{ kg})(10 \text{ m/s}^2) = 40 \text{ N}$.
I know it is about 1 yd from the tip of my outstretched arm to my forward-facing nose, but a meter is a bit bigger than a yard, so I would estimate that from fingertip to shoulder would be about 0.7 m. If I model my arm as a uniform cylinder of uniform density then the center of gravity would be at its center—0.35 m from the shoulder joint (the "hinge"). If I further refine the model of my arm, it is heavier nearer the shoulder and lighter toward the hand, so I will round the location of its center of gravity down to 0.30 m from the shoulder.
Solve: Since the arm is held horizontally $\sin \phi = 1$.

$$\tau = rF \sin \phi = (0.30 \text{ m})(40 \text{ N})(1) = 12 \text{ N} \cdot \text{m}$$

Assess: Your assumptions and estimates (and, indeed, your arm) might be different, but your answer will probably be in the same order of magnitude; you probably won't end up with an answer 10 times bigger or 10 times smaller.

P7.31. Prepare: First let's divide the object into two parts. Let's call part #1 the part to the left of the point of interest and part #2 the part to the right of the point of interest. Next using our sense of center of gravity, we know the center of mass of part #1 is at 12.5 m and the center of mass of part #2 is at +37.5 cm. We also know the mass of part #1 is one fourth the total mass of the object and the mass of part #2 is three fourths the total mass of the object. Finally, we can determine the gravitational torque of each part using any of the three expressions for torque as shown below:

Equation 7.3 $\tau = rF_\perp$ is straightforward to use because the forces are perpendicular to the position vectors, which locate the point of application of the force.

Equation 7.4 $\tau = r_\perp F$ is straightforward to use because the position vectors that locate the point of application of the forces are also the moment arms for the forces.

Equation 7.4 $\tau = rF\sin\phi$ is straightforward to use because the angles are either $90°$ or $270°$.

Solve: Using Equation 7.4 we obtain the following

$$\tau_{net} = \tau_1 + \tau_2 = m_1 gr_1\sin(90°) + m_2 gr_2\sin(-90°)$$

$$= (0.5 \text{ kg})(9.8 \text{ m/s}^2)(-0.125 \text{ m})(1) + (0.75 \text{ kg})(9.8 \text{ m/s}^2)(0.375 \text{ m})(-1) = -2.1 \text{ N} \cdot \text{m}$$

Assess: According to this answer, if released the object should rotate in a clockwise direction. Looking at the figure this is exactly what we would expect to happen. It might be easier to consider the whole rod as if acting at its center of mass and then the distance would be just 25 cm and there would be only one torque calculation.

P7.35. Prepare: Set up a coordinate system along the beams with the origin at the left end. Assume each beam is of uniform density so its own center of gravity is at its geometrical center. Then use Equation 7.14.
Solve: (a)

$$x_{cg} = \frac{x_1 m_1 + x_2 m_2}{m_1 + m_2} = \frac{(0.500 \text{ m})(10.0 \text{ kg}) + (1.00 \text{ m} + 1.00 \text{ m})(40.0 \text{ kg})}{10.0 \text{ kg} + 40.0 \text{ kg}} = 1.70 \text{ m}$$

(b) The gravitational torque on the two-beam system is the total weight acting at the center of gravity of the system:
$w = mg = (10.0 \text{ kg} + 40.0 \text{ kg})(9.80 \text{ m/s}^2) = 490 \text{ N}.$

$$\tau = rF\sin\phi = (1.70 \text{ m})(490 \text{ N})\sin 90° = 833 \text{ N} \cdot \text{m}$$

Because this torque is in the clockwise direction we report it as a negative torque: $\tau = -833 \text{ N} \cdot \text{m}$.

Assess: This problem could also be done by computing the gravitational torque individually on each beam and adding them up. The final answer would be the same. For the beams to remain in equilibrium some other object must supply an equal torque in the opposite direction.

P7.37. Prepare: A table tennis ball is a spherical shell, and we look that up in Table 7.1. The radius is half the diameter.
Solve:

$$I = \tfrac{2}{3}MR^2 = \tfrac{2}{3}(0.0027 \text{ kg})(0.020 \text{ m})^2 = 7.2 \times 10^{-7} \text{ kg} \cdot \text{m}^2$$

Assess: The answer is small, but then again, it isn't hard to start a table tennis ball rotating or stop it from doing so.
By the way, this calculation can be done in one's head without a calculator by writing the data in scientific notation and mentally keeping track of the significant figures:

$$I = \frac{2}{3}(2.7\times10^{-3} \text{ kg})(2.0\times10^{-2} \text{ m})^2 = \frac{2}{3}(27\times10^{-4} \text{ kg})(2\times10^{-2} \text{ m})^2 = \frac{27}{3}\cdot 2 \cdot 2^2 \times 10^{-4} \times 10^{-4} \text{ kg} \cdot \text{m}^2$$

$$= 9 \cdot 8 \times 10^{-8} \text{ kg} \cdot \text{m}^2 = 72 \times 10^{-8} \text{ kg} \cdot \text{m}^2 = 7.2 \times 10^{-7} \text{ kg} \cdot \text{m}^2$$

P7.41. Prepare: Treat the bicycle rim as a hoop, and use the expression given in Table 7.1 for the moment of inertia of a hoop. Manipulate this expression to obtain the mass.
Solve: The mass of the rim is determined by

$$m = I/R^2 = 0.19 \text{ kg} \cdot \text{m}^2/(0.65 \text{ m}/2)^2 = 1.8 \text{ kg}$$

Assess: Note that the units reduce to kg as expected. This amount seems a little heavy, but since we are not told what type of bicycle it is not an unreasonable amount.

P7.43. Prepare: We can calculate the angular acceleration using Equation 7.22.
Solve: Using Equation 7.22 we find
$$\tau = I\alpha = (14.0\times10^{-5}\text{ kg}\cdot\text{m}^2)(150\text{ rad/s}^2) = 6.0\times10^{-3}\text{N}\cdot\text{m}$$
Assess: This result is reasonable. The moment of inertia of the grinding wheel is small.

P7.45. Prepare: We are given the moment of inertia I and the angular acceleration α of the object. Since $\tau = I\alpha$ is the rotational analog of Newtons second law $F = ma$, we can use this relation to find the net torque on the object.
Solve: $\tau = (2.0\text{ kg}\cdot\text{m}^2)(4.0\text{ rad/s}^2) = 8.0\text{ kg}\cdot\text{m}^2/\text{s}^2 = 8.0\text{ N}\cdot\text{m}$
Assess: The units are correct and the relatively small torque magnitude is consistent with the size of the moment of inertia and the angular acceleration.

P7.49. Prepare: Equation 7.22 can be used to calculate the frictional torque once the angular acceleration is known. Using Synthesis 7.1 we can calculate the angular acceleration.
Solve: The wheel stops spinning in 12 seconds from an initial angular velocity of 0.72 revolutions per second. The initial angular velocity in radians per second is
$$\omega_i = 0.72\text{ rev/s} = (0.72\text{ rev/s})\left(\frac{2\pi\text{ rad}}{\text{rev}}\right) = 4.5\text{ rad/s}$$
The angular acceleration of the wheel is
$$\alpha = \frac{\Delta\omega}{\Delta t} = \frac{-4.5\text{ rad/s}}{12\text{ s}} = -0.38\text{ rad/s}^2$$
The only force acting on the wheel is friction. Using the results above in Equation 7.22 the *magnitude* of the frictional torque is
$$\tau = I\alpha = (0.30\text{ kg}\cdot\text{m}^2)(0.38\text{ rad/s}^2) = 0.11\text{ N}\cdot\text{m}$$
Assess: This is a relatively large amount of friction. Assuming the torque of the wheel bearings acts at a radius of around a centimeter the frictional force exerted by the bearings is over 10 N. This shows you just one of the problems you will have if you do not properly maintain your bicycle.

P7.53. Prepare: We can use the rolling constraint to find the speed of the dot and the angular speed of the tires.
Solve: (a) Since the tires are rolling without slipping the angular velocity of the tires is given by Equation 7.25.
$$\omega = \frac{v_{cm}}{R} = \frac{5.6\text{ m/s}}{0.40\text{ m}} = 14\text{ rad/s}$$
(b) The blue dot is undergoing translation and rotation. At the top of the tire, it has a translational velocity equal to the speed of the bike, and an additional velocity equal to ωR due to the rotation of the tire. See Figure 7.43. The speed of the dot at this point is
$$v = v_{cm} + \omega R = 2v_{cm} = 2(5.6\text{ m/s}) = 11\text{ m/s}$$
(c) See the diagram below.

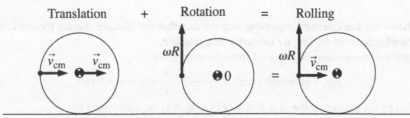

The dot has a translational velocity equal to the velocity of the center of mass of the tire in the horizontal direction. The tangential velocity of the dot is in the vertical direction and has a magnitude equal to ωR. The velocity of the dot is equal to the sum of these two vectors. Since the tire is rolling without slipping, $v_{cm} = \omega R$.

See the vector diagram below.

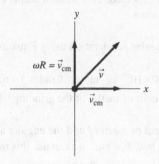

The speed of the dot is equal to

$$v = \sqrt{(v_{cm})^2 + (v_{cm})^2} = 7.9 \text{ m/s}$$

Assess: Note that the speed of the dot during the downward motion for part **(c)** is equal to the speed of the dot during the upward motion shown in the diagram.

P7.55. Prepare: The particle is moving in a circle and a motion of 2_π radians corresponds to one full rotation. Angular velocity at any given time is defined as $\omega = \Delta\theta / \Delta t$, or the slope of the angular position-versus-time graph.
Solve: (a) From $t = 0$ s to $t = 1$ s the particle rotates clockwise from the angular position $+4\pi$ rad to -2π rad. Therefore, $\Delta\theta = -2\pi - (+4\pi) = -6\pi$ rad in one sec, or $\omega = -6\pi$ rad/s. From $t = 1$ s to $t = 2$ s, $\omega = 0$ rad/s. From $t = 2$ s to $t = 4$ s the particle rotates counterclockwise from the angular position -2π rad to 0 rad. Thus $\Delta\theta = 0 - (-2\pi) = 2\pi$ rad rad and $\omega = +\pi$ rad/s.
(b)

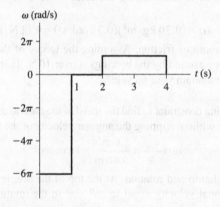

Assess: Since we take positive angular displacements as counterclockwise and negative angular displacements clockwise, we know the particle is traveling around the circle in a clockwise direction. Knowing that the slope of the angular position-versus-time plot is the angular velocity, we can establish a plot of angular velocity-versus-time.

P7.57 Prepare: Knowing the kinematic equations and the fact that the distance the car travels is some number of revolutions (or circumferences) of the tire, we can solve this problem.
Solve: The acceleration of the car may be obtained from the expression:

$$v = v_o + at \text{ which gives } a = \frac{v - v_o}{t} = \frac{v}{t} \text{ since } v_o = 0 \text{ m/s}$$

The distance traveled by the car during the time it is accelerating may be determined by:

$$2a\Delta x = v^2 - v_o^2 \text{ which gives } \Delta x = \frac{v^2 - v_o^2}{2a} = \frac{v^2}{2a} = \frac{v^2}{2(v/t)} = \frac{vt}{2}$$

Finally, the number of times the tires rotate (i.e. the number of circumferences of the tires) may be determined by:

$\Delta x = N(2\pi r) = N\pi d$ which gives $N = \dfrac{\Delta x}{\pi d} = \dfrac{(vt/2)}{\pi d} = \dfrac{vt}{2\pi d} = \dfrac{(20 \text{ m/s})(10 \text{ s})}{2[\pi(0.58 \text{ m})/\text{rotation}]} = 55$ rotations

Assess: As with many kinematics problems, we can check our work by approaching the problem in a different manner. For example, since the acceleration is constant, the distance the car travels is just the average velocity times the time of travel. This may be expressed as follows:

$\Delta x = v_{ave}t = \left(\dfrac{v+v_o}{2}\right)t = \dfrac{vt}{2}$ which is the same as the expression obtained above for the distance traveled.

Note also that the final units are in rotations and that 55 is a reasonable number of rotations for a tire in 10 seconds.

P7.63. Prepare: We can use Equation 7.21, and Table 7.1.
Solve: Refer to the figure below.

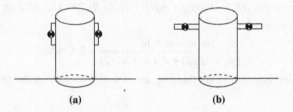

(a) The moment of inertia of the skater will be the moment of inertia of her body plus the moment of inertia of her arms. The center of mass of each arm is at her side, 20 cm from the axis of rotation. The mass of each arm is half of one eighth of the mass of her body, which is 4 kg. The mass of her body is $64 \text{ kg} - 8 \text{ kg} = 56 \text{ kg}$. With her arms at her sides, her total moment of inertia is

$$I = I_{body} + I_{arm} + I_{arm} = \tfrac{1}{2}M_{body}(R_{body})^2 + M_{arm}(R_{arm})^2 + M_{arm}(R_{arm})^2$$
$$= \tfrac{1}{2}(56 \text{ kg})(0.20 \text{ m})^2 + (4 \text{ kg})(0.20 \text{ m})^2 + (4 \text{ kg})(0.20 \text{ m})^2 = 1.4 \text{ kg} \cdot \text{m}^2$$

(b) With her arms outstretched, the center of mass of her arms is now 50 cm from the axis of rotation. Her moment of inertia is now

$$I = I_{body} + I_{arm} + I_{arm} = \tfrac{1}{2}M_{body}(R_{body})^2 + M_{arm}(R_{arm})^2 + M_{arm}(R_{arm})^2$$
$$= \tfrac{1}{2}(56 \text{ kg})(0.20 \text{ m})^2 + (4 \text{ kg})(0.50 \text{ m})^2 + (4 \text{ kg})(0.50 \text{ m})^2 = 3.1 \text{ kg} \cdot \text{m}^2$$

Her moment of inertia has increased.
Assess: Her moment of inertia with arms outstretched is almost twice as large as with them at her side. This is reasonable, since their distance from the axis of rotation is much larger.

P7.65. Prepare: This problem requires a knowledge of translational ($F_{net} = ma$) and rotational ($\tau_{net} = I\alpha$) dynamics. Notice that the counterclockwise torque is greater than the clockwise torque, hence the system will rotate counterclockwise. Let's agree to call any force that tends to accelerate the system positive and any force that tends to decelerate the system negative. Also let's agree to call the small disk M_1 and the large disk M_2.
Solve:

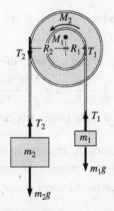

Write Newton's second law equation for m_2 and m_1 as follows:

$$m_2 g - T_2 = m_2 a_2 = m_2 R_2 \alpha \quad \text{or} \quad T_2 = m_2 g - m_2 R_2 \alpha$$

and

$$T_1 - m_1 g = m_1 a_1 = m_1 R_1 \alpha \quad \text{or} \quad T_1 = m_1 g + m_1 R_1 \alpha$$

The net torque acting on the system may be determined by

$$\tau = R_2 T_2 - R_1 T_1 = R_2 (m_2 g - m_2 \alpha R_2) - R_1 (m_1 g + m_1 \alpha R_1)$$

The moment of inertia of the system is

$$I = I_1 + I_2 = (M_1 R_1^2 / 2) + (M_2 R_2^2 / 2)$$

Knowing

$$\tau = I\alpha$$

we may combine the above to get

$$R_2 (m_2 g - m_2 \alpha R_2) - R_1 (m_1 g + m_1 \alpha R_1) = [(M_1 R_1^2 / 2) + (M_2 R_2^2 / 2)]\alpha$$

which may be solved for α to obtain

$$\alpha = \frac{(m_2 R_2 - m_1 R_1)g}{R_2^2 (m_2 + M_2 / 2) + R_1^2 (m_1 + M_1 / 2)} = 3.5 \text{ rad/s}^2$$

Assess: This angular acceleration amounts to speeding up about a half revolution per second every second. That is not an unreasonable amount.

P7.67. Prepare: Two balls connected by a rigid, massless rod are a rigid body rotating about an axis through the center of gravity. Assume that the size of the balls is small compared to 1 m. We have placed the origin of the coordinate system on the 1.0 kg ball. Since $\tau = I_{\text{about cg}} \alpha$, we need the moment of inertia and the angular acceleration to be able to calculate the required torque.

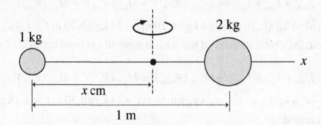

Solve: The center of gravity and the moment of inertia are

$$x_{cm} = \frac{(1.0 \text{ kg})(0 \text{ m}) + (2.0 \text{ kg})(1.0 \text{ m})}{(1.0 \text{ kg} + 2.0 \text{ kg})} = 0.667 \text{ m} \quad \text{and} \quad y_{cm} = 0 \text{ m}$$

$$I_{\text{about cm}} = \sum m_i r_i^2 = (1.0 \text{ kg})(0.667 \text{ m})^2 + (2.0 \text{ kg})(0.333 \text{ m})^2 = 0.667 \text{ kg} \cdot \text{m}^2$$

We have $\omega_f = 0 \text{ rad/s}$, $t_f - t_i = 5.0 \text{ s}$, and $\omega_i = 20 \text{ rpm} = 20(2\pi \text{ rad}/60 \text{ s}) = \frac{2}{3}\pi \text{ rad/s}$, so $\omega_f = \omega_i + \alpha(t_f - t_i)$ becomes

$$0 \text{ rad/s} = \left(\frac{2\pi}{3} \text{ rad/s}\right) + \alpha(5.0 \text{ s}) \Rightarrow \alpha = -\frac{2\pi}{15} \text{ rad/s}^2$$

Having found I and α, we can now find the torque τ that will bring the balls to a halt in 5.0 s:

$$\tau = I_{\text{about cm}} \alpha = \left(\frac{2}{3} \text{ kg} \cdot \text{m}^2\right)\left(-\frac{2\pi}{15} \text{ rad/s}^2\right) = -\frac{4\pi}{45} \text{ N} \cdot \text{m} = -0.28 \text{ N} \cdot \text{m}$$

Assess: The minus sign with the torque indicates that the torque acts clockwise.

P7.71. Prepare: This is an excellent review problem. In order to solve this problem you will need a working knowledge of rotational kinematics ($\omega_f = \omega_o + \alpha t$), moment of inertia of a cylinder ($I = MR^2 / 2$), rotational dynamics ($\tau = I\alpha$), torque ($\tau = Rf_k \sin\phi$), and kinetic friction ($f_k = \mu_k N$).

Solve: First determine an expression for the angular acceleration

$$\alpha = (\omega_f - \omega_o)/\Delta t$$

Next obtain an expression for the moment of inertias of the grindstone

$$I = MR^2/2$$

Then obtain an expression for the torque acting on the grindstone

$$\tau = I\alpha = \left(\frac{MR^2}{2}\right)\left(\frac{\omega_f - w_o}{\Delta t}\right)$$

Write a second expression for the torque in terms of the force of friction and then the normal force

$$\tau = f_k R = \mu_k N R$$

Finally, equate the last two expressions for the torque and solve for N (the force with which the man presses the knife against the grindstone).

$$N = \frac{MR(\omega_f - \omega_o)}{2\mu_k \Delta t} = \frac{(28\,\text{kg})(0.15\,\text{m})(180\,\text{rev/min} - 200\ \text{rev/min})(2\pi\,\text{rad/rev})(\text{min}/60\text{s})}{2(0.2)(10\text{s})} = 2.2\ \text{N}$$

Assess: This is a reasonable force (i.e. one that the man could easily exert and yet not grind the knife to a sliver in a matter of minutes).

Q8.3. Reason: The ladder could *not* be in static equilibrium. Consider the forces in the horizontal direction. There is a normal force exerted by the wall on the top of the ladder, but no other object (in the absence of friction) exerts a counterbalancing force on the ladder in the opposite direction. Examine Figure 8.9.

Assess: This makes sense from a common sense standpoint. If a ladder is about to slip out one tries to increase the friction at the point of contact with the floor, or to produce a horizontal component of a normal force by wedging it.

Q8.5. Reason: For divers to be stable on the board before the dive their center of gravity must be over an area of support, that is, over the board. Extending their arms moves the center of gravity over the board.

Assess: If the arms are not extended, then the center of gravity would be over the edge of the diving board when they stand on their toes with heels extended out. They would not be in static equilibrium and would topple over before getting off a good clean dive. The other option to get the center of gravity over the board (besides extending arms) is to lean forward slightly toward the board.

Q8.9. Reason: Before Carlos came along the wall also pulled on the spring with a 200 N force when Bob did, that is, there was a 200 N tension force all along the spring. When Carlos arrives he takes the place of the wall but the spring must still stretch 20 cm. The only difference is that now Carlos also moves whereas the wall didn't.

(a) 10 cm. Though the spring stretched 20 cm originally, its center moved by 10 cm and so Bob's end moved 10 cm away from (farther than) the center. In the tug-of-war the center stays still so Bob's end only moves 10 cm.

(b) 10 cm in the other direction. The total stretch under a 200 N tension must still be 20 cm.

Assess: These answers fit well with Hooke's law. In either case the 200 N tension produced a total stretch of 20 cm.

Q8.11. Reason: Assume the springs obey Hooke's law and the person is in equilibrium. If the person is heavier then the weight force is greater, so the upward force of the springs must also be greater so the sum can still be zero. To push up more, the springs must be compressed more.

Assess: These answers fit well with everyday experience.

Q8.13. Reason: The longer wire will also break at 5000 N. Because the two wires have the same diameter, the force per cross-sectional area is the same in both cases.

Assess: You see that the length of the cable is not mentioned in Table 8.2 nor in the example in the text; this means the length doesn't matter.

And the length *doesn't* matter if the diameter is truly uniform and there are no defects in the cable. In practice, however, the longer a wire or cable is the more likely it is that it will contain a defect (impurity or a chance region of slightly smaller diameter) and it is therefore more likely to fail at a slightly lower tension (at the defect) than a shorter cable or wire would.

Q8.17. Reason: Use equilibrium calculations $\Sigma\tau = 0$ around the suspended end. The weight of the rod acts at its center ($L/2$ away from the right end) and the normal force acts at L from the right end. For $\Sigma\tau = 0$ the normal force must be half the force at twice the distance, so the answer is A.
Assess: For $\Sigma F = 0$ the tension in the suspension string must also be 7.0 N. This also makes sense when computing the torques around the center of the rod.

Q8.21. Reason: As is quite clear from the figure, the force of the tendon on the lower leg is up. However, if we compute torques around a pivot point where the tendon attaches to the lower leg we see that there is a counterclockwise torque produced by the weight of the lower leg acting at the center of gravity 25 cm from the knee joint. There must be a clockwise torque to counterbalance if the lower leg is to be in equilibrium. That clockwise torque is provided by the force of the upper leg *down* on the lower leg at the knee joint.
The correct answer is B.
Assess: This may be surprising at first. It may help to visualize it as a lever with an upside down fulcrum (suspension point) where the tendon attaches to the lower leg.

Q8.23. Reason: Because $\Delta L = FL/AY$ if L is doubled then A must be doubled too to keep ΔL the same. To double A requires an increase in the diameter by a factor of $\sqrt{2}$. So the answer is B.
Assess: Doubling the diameter (an incorrect answer) would actually make the twice-as-long wire stretch less.

Q8.25. Reason: The tension in the rope is $T = mg = (20 \text{ kg})(9.80 \text{ m/s}^2) = 196$ N. Use this tension as the F in the equation below. The length of the rope is $2.5 \text{ m} + 1.0 \text{ m} = 3.5$ m.
$$\Delta L = \frac{FL}{YA} = \frac{(196 \text{ N})(3.5 \text{ m})}{(1.5 \times 10^9 \text{ N/m}^2)\pi(0.0025 \text{ m})^2} = 2.3 \text{ cm}$$
The answer is then D.
Assess: A 2.3 cm stretch of 3.5 m of rope seems reasonable.

Problems

P8.3. Prepare: Compute the torques around the bottom of the right leg of the table. The horizontal distance from there to the center of gravity of the table is $\dfrac{2.10 \text{ m}}{2} = 0.55 \text{ m} = 0.50 \text{ m}$.

Solve: Call the horizontal distance from the bottom of the right leg to the center of gravity of the man x.
$$\Sigma\tau = (56 \text{ kg})(9.8 \text{ m/s}^2)(0.50 \text{ m}) - (70 \text{ kg})(9.8 \text{ m/s}^2)x = 0 \Rightarrow x = 0.40 \text{ m}$$
The distance from the right edge of the table is now $0.55 \text{ m} - 0.40 \text{ m} = 0.15 \text{ m} = 15$ cm.

Assess: It seems likely that the table would tip if the man were closer than 15 cm to the edge.

P8.5. Prepare: Assume the pole is uniform in diameter and density. We will use the equilibrium equation $\Sigma\tau = 0$. The weight of the pole is $w = mg = (25 \text{ kg})(9.80 \text{ m/s}^2) = 245$ N.

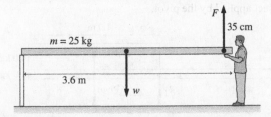

Solve: Compute the torques around the left end, where the pole rests on the fence. The weight of the pole (acting at its center of gravity) will produce a clockwise (negative) torque, and force F near the right end of the pole will produce a counterclockwise (positive) torque.

$$\Sigma\tau = (3.6\text{ m} - 0.35\text{ m})F - (1.8\text{ m})(245\text{ N})$$
$$= (3.25\text{ m})F - (1.8\text{ m})(245\text{ N})$$
$$= 0\text{ N}\cdot\text{m}$$

Now solve for F:

$$F = \frac{(1.8\text{ m})(245\text{ N})}{3.25\text{ m}} = 140\text{ N}$$

Assess: This really *is* a rest because you only have to exert a force of just over half the pole's weight, instead of the whole weight when you were carrying it. The fence is helping hold up the pole.

Think carefully about the figure and imagine moving your hands toward the fence. The upward force you would have to exert to keep $\Sigma\tau = 0$ would increase, and when you support the pole at its center of gravity the torque equation says your force is equal to the weight of the pole. At that point the fence is no longer helping (and you aren't resting) as it exerts no upward force. If you tried moving even farther toward the fence, past the center of gravity, there would be no way to keep the pole in equilibrium and it would rotate, fall, and hit the ground.

P8.7. Prepare: The rod is in rotational equilibrium, which means that $\tau_{net} = 0$. As the weight of the rod and the hanging mass pull down (the rotation of the rod is exaggerated in the figure below), the rod touches the pin at two points. The piece of the pin at the very end pushes down on the rod; the right end of the pin pushes up on the rod. To understand this, hold a pen or pencil between your thumb and forefinger, with your thumb on top (pushing down) and your forefinger underneath (pushing up).

We will calculate the torque about the left end of the rod. The downward force exerted by the pin acts through this point, so it exerts no torque. To prevent rotation, the pin's normal force $\vec{n}_{pin}$ exerts a positive torque (counterclockwise about the left end) to balance the negative torques (clockwise) of the weight of the mass and rod. The weight of the rod acts at the center of gravity.

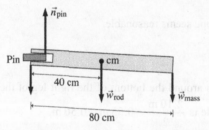

Solve:

$$\tau_{net} = 0\text{ N}\cdot\text{m} = \tau_{pin} - (0.40\text{ m})(2.0\text{ kg})(9.8\text{ m/s}^2) - (0.80\text{ m})(0.50\text{ kg})(9.8\text{ m/s}^2) \Rightarrow \tau_{pin} = 12\text{ N}\cdot\text{m}$$

Assess: We carefully chose our pivot at the left end of the rod where an unknown force acts, which contributes nothing to the torque.

P8.11. Prepare: The object balanced on the pivot is a rigid body. Since the object is balanced on the pivot, it is in both translational equilibrium and rotational equilibrium. There are three forces acting on the object: the weight $\vec{w}_1$ acting through the center of gravity of the long rod, the weight $\vec{w}_2$ acting through the center of gravity of the short rod, and the normal force $\vec{P}$ on the object applied by the pivot.

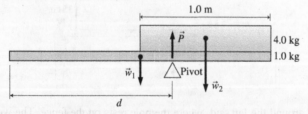

Solve: The translational equilibrium equation $(F_{net})_y = 0\text{ N}$ is

$$-w_1 - w_2 + P = 0\text{ N} \Rightarrow P = w_1 + w_2 = (1.0\text{ kg})(9.8\text{ m/s}^2) + (4.0\text{ kg})(9.8\text{ m/s}^2) = 49.0\text{ N}$$

Measuring torques about the left end, the equation for rotational equilibrium $\tau_{net} = 0$ is

$$Pd - w_1(1.0\,\text{m}) - w_2(1.5\,\text{m}) = 0\,\text{N} \cdot \text{m} \Rightarrow (49.0\,\text{N})d - (1.0\,\text{kg})(9.8\,\text{m/s}^2)(1.0\,\text{m}) - (4.0\,\text{kg})(9.8\,\text{m/s}^2)(1.5\,\text{m})$$

$$= 0\,\text{N} \Rightarrow d = 1.4\,\text{m}$$

Thus, the pivot is 1.4 m from the left end.

Assess: From geometry, the distance d is expected to be more than 1.0 m but less than 1.5 m. A value of 1.4 m is reasonable.

P8.15. Prepare: The beam is in static equilibrium. Equation 8.1 applies.
Solve: Refer to the diagram below.

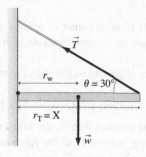

The net torque on the beam is zero. Calculating torques about the pivot, the torque equation in Equation 8.1 gives

$$r_T T_\perp - (r_w)_\perp w = r_T T \sin\theta - r_w w = 0$$

Solving for the tension,

$$T = \frac{r_w w}{r_T \sin\theta} = \frac{(0.5\,\text{m})(10\,\text{kg})(9.80\,\text{m/s}^2)}{(1.0\,\text{m})\sin 30°} = 98\,\text{N}$$

Assess: Note that using different forms of the definition of torque led to simpler calculations.

P8.17. Prepare: The critical angle is when the center of gravity is just over the edge of the cabinet.

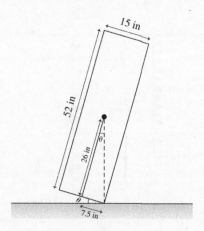

Solve: Solve the right triangle for the desired angle.

$$\theta_c = \tan^{-1}\left(\frac{7.5\,\text{in}}{26\,\text{in}}\right) = 16°$$

Assess: The file cabinet is tall, so the critical angle is small. Loading the lower drawers more would lower the center of gravity and increase the critical angle.

P8.19. Prepare: As long as the center of gravity of the board is over one of the tables, the board will not tilt.
Solve: See the figure below.

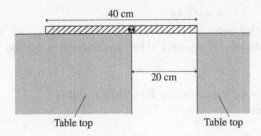

When the end of the board reaches the other table its center of gravity must still be over the first table, otherwise it will tilt. The board's center of gravity should be 20 cm from one of its ends. Assuming the board is uniform, the board must be at least 40 cm long.
Assess: For an object to be stable, the center of gravity of the object must lie over its base of support.

P8.23. Prepare: Hooke's law is given in Equation 8.3, $(F_{sp})_x = -k\Delta x$. It relates the force on a spring to the stretch; the constant of proportionality is k, the spring constant that we are asked to find.
Solve: The minus sign in Equation 8.3 simply indicates that the force and the stretch are in opposite directions (that the force is a *restoring* force); k is always positive, so we'll drop the minus sign and just use magnitudes of $(F_{sp})x$ and Δx since we would otherwise have to set up a more explicit coordinate system. See Equation 8.1.

$$k = \frac{(F_{sp})_x}{\Delta x} = \frac{25\,\text{N}}{0.030\,\text{m}} = 830\,\text{N/m}$$

Assess: This result indicates a fairly stiff spring, but certainly within reason.

P8.27. Prepare: A visual overview below shows the details, including a free-body diagram, of the problem. We will assume an ideal spring that obeys Hooke's law.

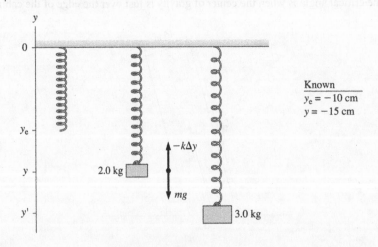

Solve: (a) The spring force on the 2.0 kg mass is $F_{sp} = -k\Delta y$. Notice that Δy is negative, so F_{sp} is positive. This force is equal to mg, because the 2.0 kg mass is at rest. We have $-k\Delta y = mg$. Solving for k:

$$k = -(mg/\Delta y) = -(2.0\,\text{kg})(9.80\,\text{m/s}^2)/(-0.15\,\text{m} - (-0.10\,\text{m})) = 392\,\text{N/m} = 390\,\text{N/m}$$

(b) Again using $-k\Delta y = mg$:

$$\Delta y = -mg/k = -(3.0\,\text{kg})(9.80\,\text{m/s}^2)/(392\,\text{N/m}) = -0.075\,\text{m}$$

$$y' - y_e = -0.075\,\text{m} \Rightarrow y' = y_e - 0.075\,\text{m} = -0.10\,\text{m} - 0.075\,\text{m} = -0.175\,\text{m} = -18\,\text{cm}$$

The length of the spring is 18 cm when a mass of 3.0 kg is attached to the spring.

Assess: The *position* of the end of the spring is negative because it is below the origin, but length must be a positive number. We expected the length to be a little more than 15 cm.

P8.29. Prepare: Assume the spring is massless and obeys Hooke's law.
Solve: The spring constant of the spring can be determined from the original stretch. The spring is at its equilibrium length before the first mass is added. Once the mass is added, assume it is in equilibrium. The net force on the mass is zero.
From Equation 8.1, $F_{sp} = mg$. Using Equation 8.3,

$$k = -\frac{F_{sp}}{\Delta y} = -\frac{mg}{\Delta y} = -\frac{(1.2 \text{ kg})(9.80 \text{ m/s}^2)}{-(0.024 \text{ m})} = 490 \text{ N/m}$$

When the new mass is added the spring stretches by

$$\Delta y = -\frac{F_{sp}}{k} = -\frac{mg}{k} = -\frac{(1.8 \text{ kg})(9.80 \text{ m/s}^2)}{490 \text{ N/m}} = -0.036 \text{ m}$$

With both masses, the spring is stretched 3.6 cm. It has stretched an additional $3.6 \text{ cm} - 2.4 \text{ cm} = 1.2 \text{ cm}$.
Assess: The stretch in the spring is always measured from its equilibrium length.

P8.33. Prepare: Equation 8.6 relates the quantities in question.
Look up Young's modulus for steel in Table 8.1: $Y_{steel} = 20 \times 10^{10} \text{ N/m}^2$.
Convert all length data to meters: $D = 1.0 \text{ cm} = 0.010 \text{ m}$, $\Delta L = 5.0 \text{ mm} = 0.0050 \text{ m}$.
Assume a circular cross section: $A = \pi r^2 = \pi(\frac{D}{2})^2 = \pi(0.0050 \text{ m})^2 = 7.85 \times 10^{-5} \text{ m}^2$.
Solve:

$$F = \frac{YA}{L}\Delta L = \frac{(20 \times 10^{10} \text{ N/m}^2)(7.85 \times 10^{-5} \text{ m}^2)}{10 \text{ m}}0.0050 \text{ m} = 7900 \text{ N}$$

This is the force required to stretch a steel cable of the given length and diameter by 5.0 mm.
Assess: A 1-cm-diameter cable is fairly substantial, so it ought to take a few thousand newtons to stretch it 5.0 mm. Notice the m^2 cancel in the numerator and so do the other m, leaving only N.

P8.35. Prepare: Equation 8.5 relates the quantities in question. Table 8.1 gives Young's modulus for steel.
Solve: (a)

$$F = \frac{YA}{L}\Delta L = \frac{(20 \times 10^{10} \text{ N/m}^2)(\pi(0.00125 \text{ m})^2)}{1.0 \text{ m}}(0.0010 \text{ m}) = 980 \text{ N}$$

(b) Assume the second wire is made of steel also. Solve Equation 8.5 for L.

$$\frac{L_2}{L_1} = \frac{\frac{Y_2 A_2}{F_2}\Delta L_2}{\frac{Y_1 A_1}{F_1}\Delta L_1}$$

The two forces are the same, the two Ys are the same, and the two ΔLs are the same. The new area is just four times the old area.

$$\frac{L_2}{L_1} = \frac{\frac{Y_2 A_2}{F_2}\Delta L_2}{\frac{Y_1 A_1}{F_1}\Delta L_1} = \frac{\frac{Y_1(4A_1)}{F_1}\Delta L_1}{\frac{Y_1 A_1}{F_1}\Delta L_1} = 4$$

So $L_2 = 4L_1 = 4(1.0 \text{ m}) = 4.0 \text{ m}$.
Assess: This ratio technique is very powerful; we did not need the answer to part (a) to do part (b).

P8.39. Prepare: Equation 8.5 relates the quantities in question; the fractional decrease in length will be $\Delta L/L$, so rearrange the equation so $\Delta L/L$ is isolated.
Look up Young's modulus for Douglas fir in Table 8.1: $Y_{\text{Douglas fir}} = 1 \times 10^{10} \text{ N/m}^2$.
The total cross section will be three times the area of one leg:

$$A_{\text{tot}} = 3(\pi r^2) = 3\left(\pi\left(\frac{D}{2}\right)^2\right) = 3\pi(0.010 \text{ m})^2 = 9.42 \times 10^{-4} \text{ m}^2$$

Compute $F = w = mg = (75 \text{ kg})(9.80 \text{ m/s}^2) = 735 \text{ N}.$

Solve:

$$\frac{\Delta L}{L} = \frac{F}{AY} = \frac{735 \text{ N}}{(9.42 \times 10^{-4} \text{ m}^2)(1 \times 10^{10} \text{ N/m}^2)} = 7.8 \times 10^{-5}$$

This is a 0.0078% change in length.

Assess: We were not given the original length of the stool legs, but regardless of the original length, they decrease in length by only a small percentage—0.0078%—because F isn't large but A is.

P8.41. Prepare: The stress is F/A. Solve for F. The maximum stress is the tensile strength; for glass Table 8.2 says the strength is $60 \times 10^6 \text{ N/m}^2$. Table 8.1 says for glass $Y = 7 \times 10^{10} \text{ N/m}^2$.

Solve: (a)

$$F = (\text{tensile strength})(\text{area}) = (60 \times 10^6 \text{ N/m}^2)(\pi (4.5 \times 10^{-6} \text{ m})^2) = 3.8 \times 10^{-3} \text{ N}$$

So the force required to break the glass fiber is only 3.8 mN.

(b) Use the force from part **(a)** to compute the stretch.

$$\Delta L = \frac{LF}{AY} = \frac{(10 \text{ m})(3.8 \times 10^{-3} \text{ N})}{(\pi (4.5 \times 10^{-6} \text{ m})^2)(7 \times 10^{10} \text{ N/m}^2)} = 8.6 \times 10^{-3} \text{ m}$$

The data in Table 8.2 was given to only one significant figure, so we report this as 9 mm.

Assess: A stretch of just under 1 cm sounds reasonable for a 10-meter fiber.

P8.43. Prepare: The beam is a rigid body of length 3.0 m and the student is a particle. $\vec{F}_1$ and $\vec{F}_2$ are the normal forces on the beam due to the supports, $\vec{w}_{\text{beam}}$ is the weight of the beam acting at the center of gravity, and $\vec{w}_{\text{student}}$ is the student's weight. The student is 1 m away from support 2.

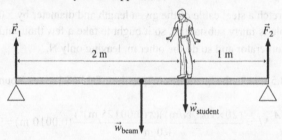

Solve: To stay in place, the beam must be in both translational equilibrium $(\vec{F}_{\text{net}} = \vec{0} \text{ N})$ and rotational equilibrium $(\tau_{\text{net}} = 0 \text{ N} \cdot \text{m})$. The first condition is

$$\Sigma F_y = -w_{\text{beam}} - w_{\text{student}} + F_1 + F_2 = 0 \text{ N} \Rightarrow F_1 + F_2 = w_{\text{beam}} + w_{\text{student}} = (100 \text{ kg} + 80 \text{ kg})(9.80 \text{ m/s}^2) = 1764 \text{ N}$$

Taking the torques about the left end of the beam, the second condition is

$$-w_{\text{beam}} (1.5 \text{ m}) - w_{\text{student}} (2.0 \text{ m}) + F_2 (3.0 \text{ m}) = 0 \text{ N} \cdot \text{m}$$

$$-(100 \text{ kg})(9.80 \text{ m/s}^2)(1.5 \text{ m}) - (80 \text{ kg})(9.80 \text{ m/s}^2)(2.0 \text{ m}) + F_2 (3.0 \text{ m}) = 0 \text{ N} \cdot \text{m} \Rightarrow F_2 = 1013 \text{ N}$$

From $F_1 + F_2 = 1764 \text{ N}$, we get $F_1 = 1764 \text{ N} - 1013 \text{ N} = 750 \text{ N}.$

Assess: To establish rotational equilibrium, the choice for the pivot is arbitrary. We can take torques about any point of interest.

P8.45. Prepare: Neglect the weight of the arm. The arm is $45°$ below horizontal which introduces $\sin 45°$ in each term, but it will cancel out.

Solve:

$$\Sigma \tau = T(14 \text{ cm})\sin 45° - (10 \text{ kg})(9.8 \text{ m/s}^2)(35 \text{ cm})\sin 45° = 0 \Rightarrow T = 860 \text{ N}$$

Assess: This answer is in the general range of the results in the example.

P8.49. Prepare: Model the beam as a rigid body. For the beam not to fall over, it must be both in translational equilibrium ($\vec{F}_{net} = \vec{0}$ N) and rotational equilibrium ($\vec{\tau}_{net} = 0$ N·m). The boy walks along the beam a distance x, measured from the left end of the beam. There are four forces acting on the beam. $\vec{F}_1$ and $\vec{F}_2$ are from the two supports, $\vec{w}_b$ is the weight of the beam, and $\vec{w}_B$ is the weight of the boy.

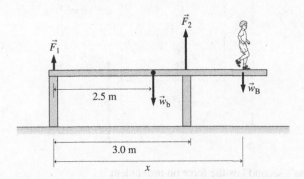

Solve: We pick our pivot point on the left end through the first support. The equation for rotational equilibrium is

$$-w_b(2.5 \text{ m}) + F_2(3.0 \text{ m}) - w_B x = 0 \text{ N·m}$$

$$-(40 \text{ kg})(9.80 \text{ m/s}^2)(2.5 \text{ m}) + F_2(3.0 \text{ m}) - (20 \text{ kg})(9.80 \text{ m/s}^2)x = 0 \text{ N·m}$$

The equation for translation equilibrium is

$$\Sigma F_y = 0 \text{ N} = F_1 + F_2 - w_b - w_B \Rightarrow F_1 + F_2 = w_b + w_B = (40 \text{ kg} + 20 \text{ kg})(9.80 \text{ m/s}^2) = 588 \text{ N}$$

Just when the boy is at the point where the beam tips, $F_1 = 0$ N. Thus $F_2 = 588$ N. With this value of F_2, we can simplify the torque equation to

$$-(40 \text{ kg})(9.80 \text{ m/s}^2)(2.5 \text{ m}) + (588 \text{ N})(3.0 \text{ m}) - (20 \text{ kg})(9.80 \text{ m/s}^2)x = 0 \text{ N·m} \Rightarrow x = 4.0 \text{ m}$$

Thus, the distance from the right end is $5.0 \text{ m} - 4.0 \text{ m} = 1.0 \text{ m}$.

P8.53. Prepare: Both springs are compressed by the same amount: $\Delta x = 1.00$ cm. Each spring obeys Hooke's law (we assume we are in the linear region of the springs) and so exerts a force back on the block with a magnitude of $F_{sp} = k\Delta x$. The net spring force will simply be the sum of the two individual spring forces.

Solve:

$$(F_{sp})_1 = k_1\Delta x = (12.0 \text{ N/cm})(1.00 \text{ cm}) = 12.0 \text{ N}$$

$$(F_{sp})_2 = k_2\Delta x = (5.4 \text{ N/cm})(1.00 \text{ cm}) = 5.4 \text{ N}$$

$$(F_{sp})_{tot} = (F_{sp})_1 + (F_{sp})_2 = 12.0 \text{ N} + 5.4 \text{ N} = 17.4 \text{ N}$$

Assess: We have purposely omitted the negative sign in Hooke's law since it only reflects the fact that the force and the stretch are in opposite directions—something we had kept in mind, but did not worry about since we only needed the magnitude of the forces.

These two springs are said to be in parallel, and they are equivalent to one spring whose spring constant is the sum of the spring constants of the two parallel springs.

P8.55. Prepare: We will model the student (S) as a particle and the spring as obeying Hooke's law. The only two forces acting on the student are his weight and the force due to the spring.

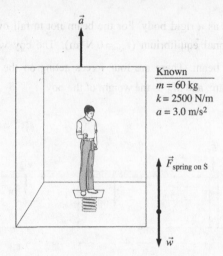

Known
$m = 60$ kg
$k = 2500$ N/m
$a = 3.0$ m/s²

$\vec{F}_{\text{spring on S}}$

$\vec{w}$

Solve: According to Newton's second law the force on the student is

$$\Sigma(F_{\text{on S}})_y = F_{\text{spring on S}} - w = ma_y \Rightarrow F_{\text{spring on S}} = w + ma_y = mg + ma_y = (60 \text{ kg})(9.80 \text{ m/s}^2 + 3.0 \text{ m/s}^2) = 768 \text{ N}$$

since $F_{\text{spring on S}} = F_{\text{S on spring}} = k\Delta y$, $k\Delta y = 768$ N. This means $\Delta y = (768 \text{ N})/(2500 \text{ N/m}) = 0.307 \text{ m} = 0.31 \text{ m}$.

P8.57. Prepare: Assume the plank is massless. Compute the torques around the pivot at the left end of the plank.

The cross section area of the rope is $A = \pi\left(\dfrac{7.0 \text{ mm}}{2}\right)^2 = 3.848 \times 10^{-5} \text{ m}^2$.

Solve: **(a)** The maximum tension the rope can support is $T = (6.0 \times 10^7 \text{ N/m}^2)(3.848 \times 10^{-5} \text{ m}^2) = 2300 \text{ N}$.

(b) Call the distance from the pivot to the machinery at maximum distance x.

$$\Sigma\tau = T(3.5 \text{ m}) - (800 \text{ kg})(9.8 \text{ m/s}^2)x = 0$$

$$x = \frac{(2300 \text{ N})(3.5 \text{ m})}{(800 \text{ kg})(9.8 \text{ m/s}^2)} = 1.0 \text{ m}$$

Assess: 1.0 m isn't very far along the plank, but the machinery is heavy. We could have moved the machinery farther out if the rope had been thicker.

P8.61. Prepare: We'll use the data from Example 8.10: $m_{\text{original}} = 70$ kg and $A_{\text{original}} = 4.8 \times 10^{-4} \text{ m}^2$.

The femur is not solid cortical bone material; we model it as a tube with an inner diameter and an outer diameter.
Look up Young's modulus for cortical bone in Table 8.1.
Solve: **(a)** Both the inner and outer diameters are increased by a factor of 10; however, the cross-sectional area of the bone material does not increase by a factor of 10. Instead, because $A = \pi R^2$, the outer cross-sectional area and the inner cross-sectional area (the "hollow" of the tube) both increase by a factor of 100. But this means that the cross-sectional area of the bone material (the difference of the outer and inner areas) also increases by a factor of 100. So the new area is $A_{\text{new}} = 100(4.8 \times 10^{-4} \text{ m}^2) = 4.8 \times 10^{-2} \text{ m}^2$.

(b) Since volume is a three-dimensional concept, if we increase each linear dimension by a factor of 10 then the volume increases by a factor of $10^3 = 1000$. We assume the density of the man is the same as before, so his mass increases by the same factor as the volume: $m_{\text{new}} = 1000(70 \text{ kg}) = 70\,000 \text{ kg}$.

(c) We follow the strategy of Example 8.10. The force compressing the femur is the man's weight, $F = mg = (70\,000 \text{ kg})(9.80 \text{ m/s}^2) = 690\,000 \text{ N}$. The resulting stress on the femur is

$$\frac{F}{A} = \frac{690\,000 \text{ N}}{4.8 \times 10^{-2} \text{ m}^2} = 1.4 \times 10^7 \text{ N/m}^2$$

A stress of 1.4×10^7 N/m^2 is 14% of the tensile strength of cortical bone given in Table 8.4.

Assess: This scaling problem illustrates clearly why animals of different sizes have different proportions. Because the volume scales with the cube of the linear dimensions and the area scales with the square of the linear dimensions then the force in F/A grows more quickly than the cross sectional area does.

PART I

FORCE AND MOTION

PptI.5. Reason: Solve $\vec{a} = \Delta\vec{v}/\Delta t$ for Δt.

$$\Delta t = \frac{\Delta \vec{v}}{\vec{a}} = \frac{20 \text{ m/s}}{6.0 \text{ m/s}^2} = 3.2 \text{ s}$$

The correct answer is C.
Assess: It would take a greyhound less time to reach top speed, but 3.2 s seems reasonable for a horse.

PptI.7. Reason: Solve $a_c = v^2/r$ for r.

$$r = \frac{v^2}{a_c} = \frac{(15 \text{ m/s})^2}{7.1 \text{ m/s}^2} = 31.7 \text{ m}$$

The correct answer is B.
Assess: From the photograph it appears 32 m is in the ballpark for the radius of the turn.

PptI.11. Reason: Solve $v_f^2 = v_i^2 + 2a\Delta x$ for Δx. The equation is simple because $v_f = 0$.

$$\Delta x = \frac{-v_i^2}{2a} = \frac{-(0.25 \text{ mm/s})^2}{2(-1.812 \text{ m/s}^2)} = 1.72\times10^{-8} \text{ m} \approx 0.02 \ \mu\text{m}$$

The answer is A.
Assess: The paramecium comes to rest in a distance much less than its own length.

PptI.13. Reason: At terminal speed the acceleration is zero, so the net force is also zero. The answer is B.
Assess: At terminal speed the magnitude of the drag force is the same as the magnitude of the gravitational force.

PptI.17. Reason: F and d are directly proportional, so if the force (due to the weight of the person) is only half, then the deflection will also be only half. The answer is B.
Assess: This makes intuitive sense since the variables are proportional.

PptI.21. Reason:
(a) For uniform circular motion there must be a net force toward the center of the circle.

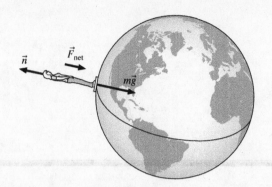

(b) The new reading is n, and the amount the reading is reduced is $mg - n$. The radius of the earth is 6.37×10^6 m. The mass of the person is $(800 \text{ N})/(9.8 \text{ m/s}^2) = 81.6$ kg. $1 \text{d} = 86400$ s

$$mg - n = \Sigma F = ma = m\left(\frac{v^2}{r}\right) = m\frac{\left(\frac{2\pi r}{\Delta t}\right)^2}{r} = m\left(\frac{2\pi}{\Delta t}\right)^2 r =$$

$$(81.6 \text{ kg})\left(\frac{2\pi}{86400 \text{ s}}\right)^2 (6.37 \times 10^6 \text{ m}) = 2.8 \text{ N}$$

Assess: The reading is only 2.8 N less than 800 N.

MOMENTUM

Q9.1. Reason: The velocities and masses vary from object to object, so there is no choice but to compute $p_x = m v_x$ for each one and then compare.

$$p_{1x} = (20 \text{ g})(1 \text{ m/s}) = 20 \text{ g} \cdot \text{m/s}$$
$$p_{2x} = (20 \text{ g})(2 \text{ m/s}) = 40 \text{ g} \cdot \text{m/s}$$
$$p_{3x} = (10 \text{ g})(2 \text{ m/s}) = 20 \text{ g} \cdot \text{m/s}$$
$$p_{4x} = (10 \text{ g})(1 \text{ m/s}) = 10 \text{ g} \cdot \text{m/s}$$
$$p_{5x} = (200 \text{ g})(0.1 \text{ m/s}) = 20 \text{ g} \cdot \text{m/s}$$

So the answer is $p_{2x} > p_{1x} = p_{3x} = p_{5x} > p_{4x}$.

Assess: The largest, most massive object did not have the greatest momentum because it was moving slower than the rest.

Q9.5. Reason: The sum of the momenta of the three pieces must be the zero vector. Since the first piece is traveling east, its momentum will have the form $(p_1, 0)$, where p_1 is a positive number. Since the second piece is traveling north, its momentum will have the form $(0, p_2)$, where p_2 is a positive number. If a third momentum is to be added to these and the result is to be $(0, 0)$, then the third momentum must be $(-p_1, -p_2)$. Since its east-west and north-south components are both negative, the momentum of the third piece must point south west and so the velocity must be south west. The answer is D.

Assess: It makes sense that the third piece would need to travel southwest. It needs a western component of momentum to cancel the eastern component of the first piece and it needs a southern component to cancel the northern component of the second piece.

Q9.7. Reason: Since both carts are stationary after the collision the final total momentum of the two-cart system is zero. By conservation of momentum, the total momentum of the system must have also been zero before the collision. Therefore the momentum of the 3 kg cart must have been the same magnitude (and opposite direction) as the momentum of the 2 kg cart. Since the momentum of the 2 kg cart was 6 kg m/s, then the speed of the 3 kg cart must have been 2 m/s.

Assess: If the carts stick together the collision is inelastic.

Q9.11. Reason: See Example 9.5. The two skaters interact with each other, but they form an isolated system because, for each skater, the upward normal force of the ice balances their downward weight force to make $\vec{F}_{\text{net}} = \vec{0}$. Thus the total momentum of the system of the two skaters will be conserved. Assume that both skaters are at rest before the push so that the total momentum before they push off is $\vec{P}_i = \vec{0}$. Consequently, the total momentum will still be $\vec{0}$ after they push off.

(a) Because the total momentum of the two-skater system is $\vec{0}$ after the push off, Megan and Jason each have momentum of the same magnitude but in the opposite direction as the other. Therefore the magnitude Δp is the same for each: $\vec{F}_{avg}\Delta t = \Delta \vec{p}$.

From the impulse-momentum theorem each experiences the same amount of impulse.

(b) They each experience the same amount of impulse because they experience the same magnitude force over the same time interval. However, over that time interval they do not experience the same acceleration. $\vec{F}_{net} = m\vec{a}$ says that since Megan and Jason experience the same force but Megan's mass is half of Jason's, then Megan's acceleration during push off will be twice Jason's. So she will have the greater speed at the end of the push off Δt.

Assess: It is important to think about both results until you are comfortable with them.

Q9.13. Reason: You do not move backward when passing the basketball because the ball-you system is not isolated: There is a net external force on the system—the friction force of the floor on your feet—to keep you from moving backward that changes the momentum of the system. If the ball-you system *is* isolated (say you are on frictionless ice), then you *do* move backward when you pass the ball.

Assess: If the friction force of the floor on you keeps you from moving backward (relative to the floor), then the law of conservation of momentum doesn't apply because the system isn't isolated. But you could then include the floor, building, and the earth in the system so it (the system) is isolated; then momentum of the system is conserved—and that means the earth does recoil ever so slightly when you pass the basketball.

Q9.15. Reason: Assume that each angular momentum is to be calculated about the axis of symmetry. It will be useful to derive a general formula for the angular momentum of a particle in uniform circular motion of radius r, calculated around the axis of symmetry. I for such a situation is mr^2, and $\omega = v/r$. Putting this all together gives the angular momentum for a particle in uniform circular motion: $L = I\omega = (mr^2)(v/r) = rmv$.

So we compute $L = rmv$ for each of the five situations.

$$L_1 = (2 \text{ m})(2 \text{ kg})(2 \text{ m/s}) = 8 \text{ kg} \cdot \text{m}^2/\text{s}$$

$$L_2 = (2 \text{ m})(3 \text{ kg})(1 \text{ m/s}) = 6 \text{ kg} \cdot \text{m}^2/\text{s}$$

$$L_3 = (2 \text{ m})(1 \text{ kg})(3 \text{ m/s}) = 6 \text{ kg} \cdot \text{m}^2/\text{s}$$

$$L_4 = (4 \text{ m})(2 \text{ kg})(1 \text{ m/s}) = 8 \text{ kg} \cdot \text{m}^2/\text{s}$$

$$L_5 = (4 \text{ m})(2 \text{ kg})(2 \text{ m/s}) = 16 \text{ kg} \cdot \text{m}^2/\text{s}$$

Finally, comparison gives $L_5 > L_1 = L_4 > L_2 = L_3$.

Assess: Since $p = mv$ the angular momentum for a particle in uniform circular motion can also be written $L = rp$, or, more generally, $L = rp_\perp$ (compare with Equation 7.10).

Q9.21. Reason: The system consisting of both blocks is isolated and so the momentum of the system is conserved.

$$\vec{P}_i = \vec{P}_f$$

$$(m)(2v_x)_i + (3m)(v_x)_i = (m + 3m)(v_x)_f$$

where the two blocks stick together to make one compound object after the collision. We want to know v_f:

$$(v_x)_f = \frac{(m)(2(v_x)_i) + (3m)(v_x)_i}{m + 3m} = \frac{5m(v_x)_i}{4m} = \frac{5(v_x)_i}{4}$$

The correct answer is D.

Assess: One could mentally confirm that the answer is reasonable by thinking of the larger block on the right. It gets a little kick from the other block and so will go a bit faster than the v_i it had before. Only answer D fits that.

Q9.23.   Reason: The initial momentum of the first block is $(p_{1x})_i = (2.5 \text{ kg})(12.0 \text{ m/s}) = 30 \text{ kg} \cdot \text{m/s}$ and the initial momentum of the second block is $(p_{2x})_i = (14 \text{ kg})(-3.4 \text{ m/s}) = -47.6 \text{ kg} \cdot \text{m/s}$. The sum of these, $-17.6 \text{ kg} \cdot \text{m/s}$, gives the total momentum, before and after the collision. After the collision, the final velocity of the two is obtained by dividing the total momentum by the total mass:

$$(v_x)_f = p_{\text{total } x}/(m_1 + m_2) = (-17.6 \text{ kg} \cdot \text{m/s})/(16.5 \text{ kg}) = -1.07 \text{ m/s}.$$

Now the final momentum of block 2 is the product of its mass and final velocity: $(p_{2x})_f = (14 \text{ kg})(-1.07 \text{ m/s}) = -15.0 \text{ kg} \cdot \text{m/s}$. The impulse of block 1 on block 2 equals the change in momentum of block 2:

$$(J_{1 \text{ on } 2})_x = \Delta p_{2x} = -15.0 \text{ kg} \cdot \text{m/s} - (-47.6 \text{ kg} \cdot \text{m/s}) = 32.6 \text{ kg} \cdot \text{m/s}.$$

Now impulse equals the area under a force versus time graph. So we need to choose the graph whose area is about $32.6 \text{ kg} \cdot \text{m/s}$. The graphs are approximately triangular, so each has an area that is approximately half its base times its height. Choice D looks the best because the height is 3000 N and the base is about 25 ms for an area of about:

$$A = \frac{1}{2}(25 \times 10^{-3} \text{ s})(3000 \text{ N}) = 37.5 \text{ N} \cdot \text{s} = 37.5 \text{ kg} \cdot \text{m/s}.$$

The other graphs differ from this by a factor of 10 or more.

Assess: Even though we can't determine the value of the force on block 2 at each time, we can find the area under the force curve which is just the impulse.

Q9.27.   Reason: The two-disk system is isolated, so the angular momentum is conserved: $L_f = L_i$. The friction force is not an external force on the system, but within the system. The moment of inertia of the identical disks is the same: call it I.

$$I\omega_1 + I\omega_2 = I(\omega_1 + \omega_2) = I_f \omega_f$$

Where I_f is the combined moment of inertia of the disks stuck together; it equals twice the moment of inertia of one disk.

$$I(\omega_1 + \omega_2) = 2I\omega_f \Rightarrow \omega_f = \tfrac{1}{2}(\omega_1 + \omega_2) = \tfrac{1}{2}(30 \text{ rpm} + 20 \text{ rpm}) = 25 \text{ rpm}$$

So the answer is C.

Assess: Intuition would have guided us to this answer. The faster disk speeds up the slower one and the slower one slows down the faster one.

Problems

P9.1.   Prepare: Model the bicycle and its rider as a particle. Also model the car as a particle. We will use Equations 9.7 for momentum.

Solve: From the definition of momentum,

$$p_{\text{car}} = p_{\text{bicycle}} \Rightarrow m_{\text{car}} v_{\text{car}} = m_{\text{bicycle}} v_{\text{bicycle}} \Rightarrow v_{\text{bicycle}} = \frac{m_{\text{car}}}{m_{\text{bicycle}}} v_{\text{car}} = \left(\frac{1500 \text{ kg}}{100 \text{ kg}} \right)(1.0 \text{ m/s}) = 15 \text{ m/s}$$

Assess: This is a reasonable speed. This problem shows the importance of mass in comparing two momenta.

P9.5.   Prepare: We use the equation $J = \Delta p$.

Solve: The initial momentum of sled and rider is $(p_x)_i = m(v_x)_i = (80 \text{ kg})(4.0 \text{ m/s}) = 320 \text{ kg} \cdot \text{m/s}$ and the final momentum of sled and rider is $(p_x)_f = m(v_x)_f = (80 \text{ kg})(3.0 \text{ m/s}) = 240 \text{ kg} \cdot \text{m/s}$. So the impulse is given by

$$J_x = (p_x)_f - (p_x)_i = 320 \text{ kg} \cdot \text{m/s} - 240 \text{ kg} \cdot \text{m/s} = 80 \text{ kg} \cdot \text{m/s}$$

Assess: This is a reasonable impulse. It could result, for example, from a force of around nine pounds for two seconds.

P9.7. **Prepare:** Model the object as a particle and its interaction with the force as a collision. We will use Equations 9.1 and 9.9. Because $p = mv$, so $v = p/m$.

Solve: (a) Using the equations

$$(p_x)_f = (p_x)_i + J_x$$

$$J_x = \text{area under the force curve} \Rightarrow (v_x)_f = (1.0 \text{ m/s}) + \frac{1}{2.0 \text{ kg}}(\text{area under the force curve})$$

$$= (1.0 \text{ m/s}) + \frac{1}{2.0 \text{ kg}}(1.0 \text{ N} \cdot \text{s}) = 1.5 \text{ m/s}$$

The force is to the right.

(b) Likewise,

$$(v_x)_f = (1.0 \text{ m/s}) + \left(\frac{1}{2.0 \text{ kg}}\right)(\text{area under the force curve}) = (1.0 \text{ m/s}) + \left(\frac{1}{2.0 \text{ kg}}\right)(-1.0 \text{ N} \cdot \text{s}) = 0.5 \text{ m/s}$$

This force is also to the right.

Assess: For an object with positive velocity, a negative impulse slows down an object and a positive impulse increases speed. The opposite is true for an object with negative velocity.

P9.9. **Prepare:** From Equations 9.5, 9.2, and 9.8, Newton's second law can be profitably rewritten as

$$\vec{F}_{avg} = \frac{\Delta \vec{p}}{\Delta t}$$

Solve: This allows us to find the force on the child and sled.

$$(F_x)_{ave} = \frac{\Delta p_x}{\Delta t} = \frac{m(v_x)_f - m(v_x)_i}{\Delta t} = \frac{m((v_x)_f - (v_x)_i)}{\Delta t} = \frac{(35 \text{ kg})(0 \text{ m/s} - 1.5 \text{ m/s})}{0.50 \text{ s}} = -105 \text{ N} \approx -110 \text{ N}$$

where the negative sign indicates that the force is in the direction opposite the original motion, as stated in the problem. So the *amount* (magnitude) of the average force you need to exert is 110 N.

Assess: This result is neither too large nor too small. In some collision problems Δt is quite a bit shorter and so the force is correspondingly larger.

P9.15. **Prepare:** This is a problem with no external forces so we can use the law of conservation of momentum.

Solve: The total momentum before the bullet hits the block equals the total momentum after the bullet passes through the block so we can write

$$m_b(v_b)_i + m_{bl}(v_{bl})_i = m_b(v_b)_f + m_{bl}(v_{bl})_f \Rightarrow$$

$$(3.0 \times 10^{-3} \text{ kg})(500 \text{ m/s}) + (2.7 \text{ kg})(0 \text{ m/s}) = (3.0 \times 10^{-3} \text{ kg})(220 \text{ m/s}) + (2.7 \text{ kg})(v_{bl})_f.$$

We can solve for the final speed of the block: $(v_{bl})_f = 0.31 \text{ m/s}$.

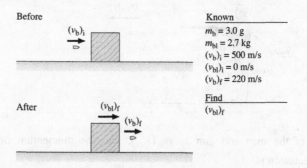

Assess: This is reasonable since the block is about one thousand times more massive than the bullet and its change in speed is about one thousand times less.

P9.17. Prepare: We will choose car + gravel to be our system. The initial x-velocity of the car is 2 m/s and that of the gravel is 0 m/s. To find the final x-velocity of the system, we will apply the momentum conservation Equation 9.15.

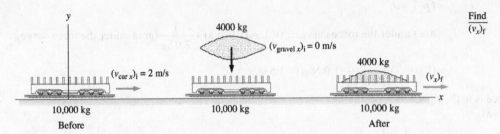

Solve: There are no *external* forces on the car + gravel system, so the horizontal momentum is conserved. This means $(p_x)_f = (p_x)_i$. Hence,

$$(10\,000\ \text{kg} + 4000\ \text{kg})(v_x)_f = (10\,000\ \text{kg})(2.0\ \text{m/s}) + (4000\ \text{kg})(0.0\ \text{m/s}) \Rightarrow (v_x)_f = 1.4\ \text{m/s}$$

Assess: The motion of railroad has to be on a level track for conservation of linear momentum to hold. As we would have expected, the final speed is smaller than the initial speed.

P9.19. Prepare: The ice is frictionless. Considering the hunter and bullet as the system, momentum will be conserved because there are no external forces acting in the horizontal direction. Assume the hunter shoots to our right.

Solve: The momentum of the system before the man fires the gun is zero because everything is at rest. The momentum of the entire system must also be zero after the man fires the gun since momentum is conserved in the system. See the diagram below.

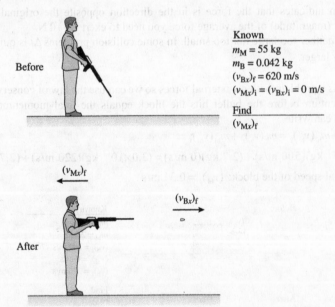

Known
$m_M = 55$ kg
$m_B = 0.042$ kg
$(v_{Bx})_f = 620$ m/s
$(v_{Mx})_i = (v_{Bx})_i = 0$ m/s

Find
$(v_{Mx})_f$

Writing the momentum of the man and gun as $m_M(v_{Mx})$, and the momentum of the bullet as $m_B(v_{Bx})$, the momentum conservation equation is

$$m_M(v_{Mx})_f + m_B(v_{Bx})_f = m_M(v_{Mx})_i + m_B(v_{Bx})_i = 0$$

Solving for $(v_{Mx})_f$,

$$(v_{Mx})_f = -\left(\frac{m_B}{m_M}\right)(v_{Bx})_f = -\left(\frac{0.042\ \text{kg}}{55\ \text{kg}}\right)(620\ \text{m/s}) = -0.47\ \text{m/s}$$

The man moves to the left with a speed of 0.47 m/s.

The man moves to the left with a speed of 0.47 m/s.

Assess: This result seems reasonable. Though the bullet has a high speed, the mass of the bullet is much smaller than the mass of the man.

P9.23. Prepare: Even though this is an inelastic collision, momentum is still conserved during the short collision if we choose the system to be spitball plus carton. Let SB stand for the spitball, CTN the carton, and BOTH be the combined object after impact (we assume the spitball sticks to the carton). We are given $m_{SB} = 0.0030$ kg, $m_{CTN} = 0.020$ kg, and $(v_{BOTHx})_f = 0.30$ m/s.

Solve:

$$(P_x)_i = (P_x)_f$$

$$(p_{SBx})_i + (p_{CTNx})_i = (p_{BOTHx})_f$$

$$m_{SB}(v_{SBx})_i + m_{CTN}(v_{CTNx})_i = (m_{SB} + m_{CTN})(v_{BOTHx})_f$$

We want to know $(v_{SBx})_i$ so we solve for it. Also recall that $(v_{CTNx})_i = 0$ m/s so the last term in the following numerator drops out.

$$(v_{SBx})_i = \frac{(m_{SB} + m_{CTN})(v_{BOTHx})_f - m_{CTN}(v_{CTNx})_i}{m_{SB}} = \frac{(0.0030\,\text{kg} + 0.020\,\text{kg})(0.30\,\text{m/s})}{0.0030\,\text{kg}} = 2.3\,\text{m/s}$$

Assess: The answer of 2.3 m/s is certainly within the capability of an expert spitballer.

P9.27. Prepare: Ignore the force of gravity since we are examining such a short time interval. Then we can use the law of conservation of momentum in both the x and y directions. Just before it gets hit the coin has zero velocity. The bullet will deflect somewhat downward. Use subscript b for the bullet and c for the coin.

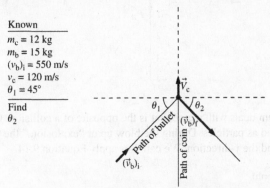

Known

$m_c = 12$ kg
$m_b = 15$ kg
$(v_b)_i = 550$ m/s
$v_c = 120$ m/s
$\theta_1 = 45°$

Find

θ_2

Solve: Using both directions, we'll get two equations in two unknowns. If we are clever we can divide the two equations to get what we want. First start with the x-direction; only the bullet moves in the x-direction.

$$\Sigma(p_x)_i = \Sigma(p_x)_i$$

$$m_b(v_b)_i \cos\theta_1 = m_b(v_b)_f \cos\theta_2 \Rightarrow$$

$$(v_b)_f \cos\theta_2 = \frac{m_b(v_b)_i \cos\theta_1}{m_b}$$

Now apply conservation of momentum in the y-direction with the initial momentum of the coin being zero.

$$\Sigma(p_y)_i = \Sigma(p_y)_i$$

$$m_b(v_b)_i \sin\theta_1 = m_b(v_b)_f \sin\theta_2 + m_c v_c \Rightarrow$$

$$(v_b)_f \sin\theta_2 = \frac{m_b(v_b)_i \sin\theta_1 - m_c v_c}{m_b}$$

Now the clever step: divide the equations so one of the unknowns (the one we don't want) cancels out.

$$\frac{(v_b)_f \sin\theta_2}{(v_b)_f \cos\theta_2} = \frac{\dfrac{m_b(v_b)_i \sin\theta_1 - m_c v_c}{m_b}}{\dfrac{m_b(v_b)_i \cos\theta_1}{m_b}}$$

The final velocity of the bullet cancels on the left side, as does the mass of the bullet in each denominator on the right.

$$\tan\theta_2 = \frac{m_b(v_b)_i \sin\theta_1 - m_c v_c}{m_b(v_b)_i \cos\theta_1} = \tan\theta_1 - \frac{m_c v_c}{m_b(v_b)_i \cos\theta_1}$$

Now plug in all the values.

$$\tan\theta_2 = \tan 45° - \frac{(12\text{ g})(120\text{ m/s})}{(15\text{ g})(550\text{ m/s})\cos 45°} \Rightarrow \theta_2 = 37°$$

The bullet ricochets away from the collision at $37°$ below the horizontal.

Assess: We expected the angle to be between 0 and 45 degrees below the horizontal.

P9.29. Prepare: We assume that the momentum is conserved in the collision.
Solve: The conservation of momentum Equation 9.14 yields

$$(p_{1x})_f + (p_{2x})_f = (p_{1x})_i + (p_{2x})_i \Rightarrow (p_{1x})_f + 0\text{ kg}\cdot\text{m/s} = 2\text{ kg}\cdot\text{m/s} - 4\text{ kg}\cdot\text{m/s} \Rightarrow (p_{1x})_f = -2\text{ kg}\cdot\text{m/s}$$

$$(p_{1y})_f + (p_{2y})_f = (p_{1y})_i + (p_{2y})_i \Rightarrow (p_{1y})_f - 1\text{ kg}\cdot\text{m/s} = 2\text{ kg}\cdot\text{m/s} + 1\text{ kg}\cdot\text{m/s} \Rightarrow (p_{1y})_f = 4\text{ kg}\cdot\text{m/s}$$

The final momentum vector of particle 1 that has the above components is shown below.

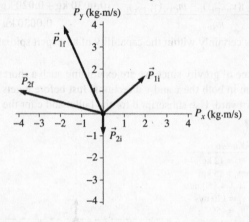

P9.31. Prepare: This problem deals with a case that is the opposite of a collision. Our system is comprised of three coconut pieces that are modeled as particles. During the blow up or "explosion," the total momentum of the system is conserved in the x-direction and the y-direction. We can thus apply Equation 9.14.

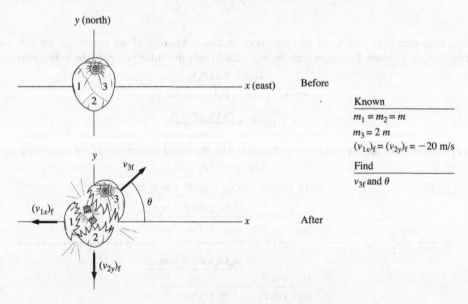

Solve: The initial momentum is zero. From $(p_x)_f = (p_x)_i$ we get

$$+m_1(v_{1x})_f + m_3(v_{3f})\cos\theta = 0 \text{ kg m/s} \Rightarrow (v_{3f})\cos\theta = \frac{-m_1(v_{fx})_1}{m_3} = \frac{-m(-20 \text{ m/s})}{2m} = 10 \text{ m/s}$$

From $(p_y)_f = (p_y)_i$, we get

$$+m_2(v_{2y})_f + m_3(v_{3f})\sin\theta = 0 \text{ kg m/s} \Rightarrow (v_{3f})\sin\theta = \frac{-m_2(v_{fy})_2}{m_3} = \frac{-m(-20 \text{ m/s})}{2m} = 10 \text{ m/s}$$

$$\Rightarrow (v_{3f}) = \sqrt{(10 \text{ m/s})^2 + (10 \text{ m/s})^2} = 14 \text{ m/s} \qquad \theta = \tan^{-1}(1) = 45°$$

Assess: The obtained speed of the third piece is of similar order of magnitude as the other two pieces, which is physically reasonable.

P9.35. **Prepare:** The disk is a rotating rigid body. The angular velocity ω is 600 rpm = $600 \times 2\pi/60$ rad/s = 20π rad/s. rad/s. From Table 7.1, the moment of inertial of the disk about its center is (1/2) MR^2, which can be used with $L = I\omega$ to find the angular momentum.
Solve:

$$I = \frac{1}{2}MR^2 = \frac{1}{2}(2.0 \text{ kg})(0.020 \text{ m})^2 = 4.0 \times 10^{-4} \text{ kg} \cdot \text{m}^2$$

Thus, $L = I\omega = (4.0 \times 10^{-4} \cdot \text{kg m}^2)(20\pi \text{ rad/s}) = 0.025 \text{ kg m}^2/\text{s}$.

Assess: For objects of size ~1 m and mass ~1 kg spinning at ~1 rad/s, an answer on order of ~1 kg· m²/s is reasonable.

P9.37. Prepare: We neglect any small frictional torque the ice may exert on the skater and apply the law of conservation of angular momentum.

$$L_i = L_f$$
$$I_i\omega_i = I_f\omega_f$$

Even though the data for $\omega_i = 5.0$ rev/s is not in SI units, it's okay because we are asked for the answer in the same units. We are also given $I_i = 0.80 \text{ kg} \cdot \text{m}^2$ and $I_f = 3.2 \text{ kg} \cdot \text{m}^2$.
Solve:

$$\omega_f = \frac{I_i\omega_i}{I_f} = \frac{(0.80 \text{ kg} \cdot \text{m}^2)(5.0 \text{ rev/s})}{3.2 \text{ kg} \cdot \text{m}^2} = 1.25 \text{ rev/s} \approx 1.3 \text{ rev/s}$$

Assess: I increased by a factor of 4, so we expect ω to decrease by a factor of 4.

P9.41. Prepare: Model the ball as a particle that is subjected to an impulse when it is in contact with the floor. We will also use constant-acceleration kinematic equations.

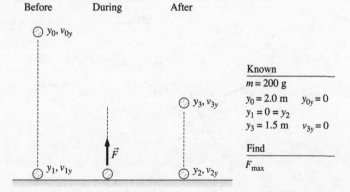

Solve: To find the ball's velocity just before and after it hits the floor:

$$v_{1y}^2 = v_{0y}^2 + 2a_y(y_1 - y_0) = 0 \text{ m}^2/\text{s}^2 + 2(-9.8 \text{ m/s}^2)(0 - 2.0 \text{ m}) \Rightarrow v_{1y} = -6.261 \text{ m/s}$$

$$v_{3y}^2 = v_{2y}^2 + 2a_y(y_3 - y_2) \Rightarrow 0 \text{ m}^2/\text{s}^2 = v_{2y}^2 + 2(-9.8 \text{ m/s}^2)(1.5 \text{ m} - 0 \text{ m}) \Rightarrow v_{2y} = 5.422 \text{ m/s}$$

The force exerted by the floor on the ball can be found from the impulse-momentum theorem:

$$J_y = \text{area under the force curve} = \Delta p_y = mv_{2y} - mv_{1y}$$

or

$$\frac{1}{2}F_{max}\Delta t = mv_{2y} - mv_{1y}$$

so that

$$F_{max} = \frac{2m(v_{2y} - v_{1y})}{\Delta t} = \frac{2(0.20 \text{ kg})(5.42 \text{ m/s} - (-6.26 \text{ m/s}))}{5.0 \times 10^{-3} \text{ s}} = 930 \text{ N}$$

Assess: A force of 930 N exerted by the floor is typical of such collisions.

P9.45.   Prepare: To find the impulse delivered by the bat to the ball, we need to know the change in the ball's momentum and use $J = \Delta p$. Since the direction of the ball changes, we need to use vector components. The x-component of the ball's final velocity is

$$(v_x)_f = (-55 \text{ m/s})\cos(25°) = -49.8 \text{ m/s} \text{ and the } y\text{-component is } (v_y)_f = (-55 \text{ m/s})\cos(25°) = -23.2 \text{ m/s}$$

Solve: The initial velocity of the ball is $(v_x)_i = 35 \text{ m/s}$ and its initial momentum is obtained by multiplying by the mass of the ball:

$$(p_x)_i = (0.140 \text{ kg})(35 \text{ m/s}) = 4.90 \text{ kg} \cdot \text{m/s} \qquad (p_y)_i = 0 \text{ kg} \cdot \text{m/s}$$

The initial final momentum is the final velocity of the ball times its mass:

$$(p_x)_f = (0.140 \text{ kg})(-49.8 \text{ m/s}) = -6.97 \text{ kg} \cdot \text{m/s}$$

$$(p_y)_f = (0.140 \text{ kg})(23.2 \text{ m/s}) = 3.25 \text{ kg} \cdot \text{m/s}$$

Finally, the impulse on the ball equals the change in the ball's momentum:

$$J_x = (p_x)_f - (p_x)_i = -11.9 \text{ kg} \cdot \text{m/s} \qquad J_y = (p_y)_f - (p_y)_i = 3.25 \text{ kg} \cdot \text{m/s}$$

The magnitude of the impulse can be obtained from the Pythagorean theorem: $J = 12 \text{ kg} \cdot \text{m/s}$ and we can find the angle, θ, above the horizontal using inverse tangent: $\theta = \tan^{-1}(3.25/11.9) = 15°$. The direction is to the left and 15° above the horizontal.

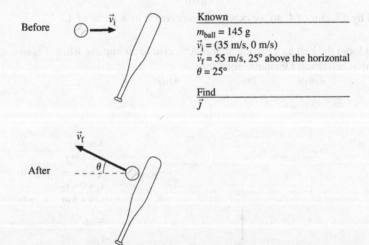

Before $\vec{v}_i$

Known

$m_{ball} = 145 \text{ g}$
$\vec{v}_i = (35 \text{ m/s}, 0 \text{ m/s})$
$\vec{v}_f = 55 \text{ m/s}, 25°$ above the horizontal
$\theta = 25°$

Find

$\vec{J}$

$\vec{v}_f$

After θ

Assess: The angle 15° makes sense because the ball comes in at 0° with the horizontal and leaves the bat at 25° above the horizontal. We expect the force, and therefore the impulse, exerted by the bat to have an angle intermediate to these two.

P9.47. Prepare: We are asked for the impulse given to the pollen grain; impulse is defined in Equation 9.1: $J = F_{avg}\Delta t$. We are given that $\Delta t = 3.0 \times 10^{-4}$ s, but we note that since we are not given any velocities, we will not use momentum or the impulse-momentum theorem. With that approach eliminated, how will we find F_{avg}? Given $m = 1.0 \times 10^{-10}$ kg and $a = 2.5 \times 10^4$ m/s^2 guides us to use Newton's second law to find F_{avg} (assuming a is constant over Δt). The F_{avg} in the impulse equation is the same as the F_{net} in Newton's law because we are ignoring any other forces on the grain.

$$F_{avg} = F_{net} = ma = (1.0 \times 10^{-10} \text{ kg})(2.5 \times 10^4 \text{ m/s}^2) = 2.5 \times 10^{-6} \text{ N}$$

Solve:

$$J = F_{avg}\Delta t = (2.5 \times 10^{-6} \text{ N})(3.0 \times 10^{-4} \text{ s}) = 7.5 \times 10^{-10} \text{ N} \cdot \text{s} = 7.5 \times 10^{-10} \text{ kg} \cdot \text{m/s}$$

Assess: This is certainly a small impulse, but the pollen grains have such small mass that the impulse is sufficient to give them the stated acceleration.

P9.49. Prepare: Let the system be ball + racket. During the collision of the ball and the racket, momentum is conserved because all external interactions are insignificantly small. We will also use the momentum-impulse theorem.

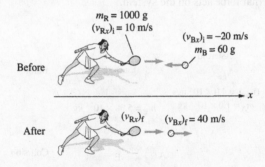

Solve: (a) The conservation of momentum equation $(p_x)_f = (p_x)_i$ is

$$m_R(v_{Rx})_f + m_B(v_{Bx})_f = m_R(v_{Rx})_i + m_B(v_{Bx})_i$$

$$(1.0 \text{ kg})(v_{Rx})_f + (0.06 \text{ kg})(40 \text{ m/s}) = (1.0 \text{ kg})(10 \text{ m/s}) + (0.06 \text{ kg})(-20 \text{ m/s}) \Rightarrow (v_{Rx})_f = 6.4 \text{ m/s}$$

(b) The impulse on the ball is calculated from $(p_{Bx})_f = (p_{Bx})_i + J_x$ as follows:

$$(0.06 \text{ kg})(40 \text{ m/s}) = (0.06 \text{ kg})(-20 \text{ m/s}) + J_x \Rightarrow J_x = 3.6 \text{ N} \cdot \text{s} = F_{avg}\Delta t$$

$$\Rightarrow F_{avg} = \frac{3.6 \text{ Ns}}{10 \text{ ms}} = 360 \text{ N}$$

Assess: Let us now compare this force with the ball's weight $w_B = m_B g = (0.06 \text{ kg})(9.8 \text{ m/s}^2) = 0.588$ N. Thus, $F_{avg} = 610 \, w_B$. This is a significant force and is reasonable because the impulse due to this force changes the direction as well as the speed of the ball from approximately 45 mph to 90 mph.

P9.51. Prepare: We will define our system to be Dan + skateboard. The system has nonzero initial momentum p_{ix}. As Dan (D) jumps backward off the gliding skateboard (S), the skateboard will move forward in such a way that the final total momentum of the system is equal to the initial momentum$x$. This conservation of momentum occurs because $\vec{F}_{ext} = \vec{0}$ on the system.

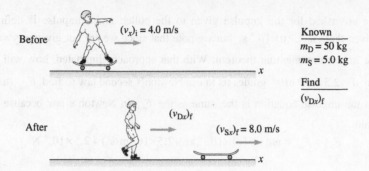

Before $(v_x)_i = 4.0$ m/s

Known
$m_D = 50$ kg
$m_S = 5.0$ kg

Find
$(v_{Dx})_f$

After $(v_{Dx})_f$ $(v_{Sx})_f = 8.0$ m/s

Solve: We have $m_S(v_{Sx})_f + m_D(v_{Dx})_f = (m_S + m_D)(v_x)_i$. Hence,

$$(5.0 \text{ kg})(8.0 \text{ m/s}) + (50 \text{ kg})(v_{Dx})_f = (5.0 \text{ kg} + 50 \text{ kg})(4.0 \text{ m/s}) \Rightarrow (v_{Dx})_f = 3.6 \text{ m/s}$$

Assess: A speed of 3.6 m/s or 8 mph is reasonable.

P9.57. Prepare: Model the earth (E) and the asteroid (A) as particles. Earth + asteroid is our system. Since the two stick together during the collision, this is a case of a perfectly inelastic collision. Momentum is conserved in the collision since no significant external force acts on the system.

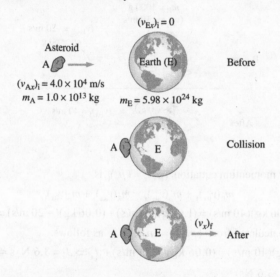

$(v_{Ex})_i = 0$

Asteroid
A

Earth (E)   Before

$(v_{Ax})_i = 4.0 \times 10^4$ m/s
$m_A = 1.0 \times 10^{13}$ kg $m_E = 5.98 \times 10^{24}$ kg

A   E   Collision

A   E $(v_x)_f$ After

Solve: (a) The conservation of momentum equation $(p_x)_f = (p_x)_i$ is

$$m_A(v_{Ax})_i + m_E(v_{Ex})_i = (m_A + m_E)(v_x)_f$$

$$\Rightarrow (1.0 \times 10^{13} \text{ kg})(4 \times 10^4 \text{ m/s}) + 0 \text{ kg m/s} = (1.0 \times 10^{13} \text{ kg} + 5.98 \times 10^{24} \text{ kg})(v_x)_f \Rightarrow (v_x)_f = 6.7 \times 10^{-8} \text{ m/s}$$

(b) The speed of the earth going around the sun is

$$v_E = \frac{2\pi r}{T} = \frac{2\pi (1.50 \times 10^{11} \text{ m})}{3.15 \times 10^7 \text{ s}} = 3.0 \times 10^4 \text{ m/s}$$

Hence, $(v_x)_f / v_E = 2.2 \times 10^{-12} = 2.2 \times 10^{-10}\%$.

Assess: The earth's recoil speed is insignificant compared to its orbital speed because of its large mass.

P9.59. Prepare: This problem deals with a case that is the opposite of a collision. The two ice skaters, heavier and lighter, will be modeled as particles. The skaters (or particles) move apart after pushing off against each other. During the "explosion," the total momentum of the system is conserved.

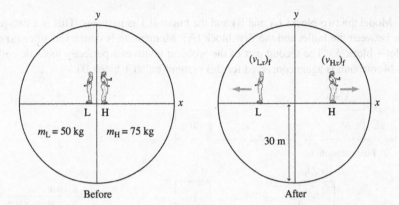

Before   After

Solve: The initial momentum is zero. Thus the conservation of momentum equation $(p_x)_f = (p_x)_i$ is

$$m_H(v_{Hx})_f + m_L(v_{Lx})_f = 0 \text{ kg} \cdot \text{m/s} \Rightarrow (75 \text{ kg})(v_{Hx})_f + (50 \text{ kg})(v_{Lx})_f = 0 \text{ kg} \cdot \text{m/s}$$

Using the observation that the heavier skater takes 20 s to cover a distance of 30 m, we find $(v_{Hx})_f = 30 \text{ m}/20 \text{ s} = 1.5 \text{ m/s}$. Thus,

$$(75 \text{ kg})(1.5 \text{ m/s}) + (50 \text{ kg})(v_{Lx})_f = 0 \text{ kg} \cdot \text{m/s} \Rightarrow (v_{Lx})_f = -2.25 \text{ m/s}$$

Thus, the time for the lighter skater to reach the edge is

$$\frac{30 \text{ m}}{(v_{Lx})_f} = \frac{30 \text{ m}}{2.25 \text{ m/s}} = 13 \text{ s}$$

Assess: Conservation of momentum leads to a higher speed for the lighter skater, and hence a shorter time to reach the edge of the ice rink. A time of 13 s at the speed of 2.3 m/s is reasonable.

P9.63. Prepare: This is a straightforward problem if done one step at a time. First, use conservation of momentum in the x direction to find the horizontal speed with which the block leaves the table. Then use kinematic equations to find how far it will go as a projectile while falling 75 cm. Ignore air resistance. Use lower case b for the bullet and upper case B for the block.

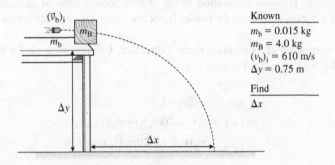

Known
$m_b = 0.015$ kg
$m_B = 4.0$ kg
$(v_b)_i = 610$ m/s
$\Delta y = 0.75$ m

Find
Δx

Solve: For the inelastic collision in the x direction:

$$m_b(v_b)_i = (m_b + m_B)v_f$$

$$v_f = \frac{m_b(v_b)_i}{m_b + m_B} = \frac{(0.015 \text{ kg})(610 \text{ m/s})}{0.015 \text{ kg} + 4.0 \text{ kg}} = 2.28 \text{ m/s}$$

Use kinematic equations on the projectile motion. First determine how long it takes to fall 75 cm from rest.

$$\Delta y = \tfrac{1}{2}g(\Delta t)^2 \Rightarrow \Delta t = \sqrt{\frac{2\Delta y}{g}} = \sqrt{\frac{2(0.75 \text{ m})}{9.8 \text{ m/s}^2}} = 0.391 \text{ s}$$

Now compute how far the projectile will travel horizontally in that time.

$$\Delta x = v\Delta t = (2.28 \text{ m/s})(0.391 \text{ s}) = 0.89 \text{ m}$$

Assess: This seems like a reasonable distance to go horizontally while falling 0.75 m.

P9.65. Prepare: Model the two blocks (A and B) and the bullet (L) as particles. This is a two-part problem. First, we have a collision between the bullet and the first block (A). Momentum is conserved since no external force acts on the system (bullet + block A). The second part of the problem involves a perfectly inelastic collision between the bullet and block B. Momentum is again conserved for this system (bullet + block B).

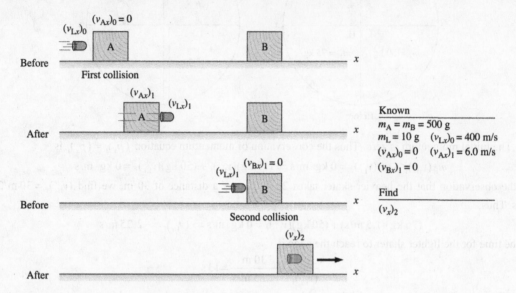

Solve: For the first collision the equation $(p_x)_f = (p_x)_i$ is

$$m_L(v_{Lx})_1 + m_A(v_{Ax})_1 = m_L(v_{Lx})_0 + m_A(v_{Ax})_0$$

$$\Rightarrow (0.01\,\text{kg})(v_{Lx})_1 + (0.500\,\text{kg})(6\,\text{m/s}) = (0.01\,\text{kg})(400\,\text{m/s}) + 0\,\text{kg m/s} \Rightarrow (v_{Lx})_1 = 100\,\text{m/s}$$

The bullet emerges from the first block at 100 m/s. For the second collision the equation $(p_x)_f = (p_x)_i$ is

$$(m_L + m_B)(v_x)_2 = m_L(v_{Lx})_1 \Rightarrow (0.01\,\text{kg} + 0.5\,\text{kg})(v_x)_2 = (0.01\,\text{kg})(100\,\text{m/s}) \Rightarrow (v_x)_2 = 2.0\,\text{m/s}$$

Assess: This problem invloves repeated application of the law of conservation of momentum. Also note that the actual value of 2 m for the separation between the blocks is not necessary for our calculations.

P9.69. Prepare: Use conservation of momentum in the x direction. Use lower case b for the ball and upper case B for the bottle. $m_b = 0.15$ kg, and $m_B = 2.0$ kg

Solve:

$$\Sigma p_i = \Sigma p_f$$

$$m_b(v_b)_i = m_b(-0.20(v_b)_i) + m_B(v_B)_f$$

$$(v_B)_f = \frac{m_b(v_b)_i + 0.20 m_b(v_b)_i}{m_B}$$

$$\frac{(v_B)_f}{(v_b)_i} = \frac{(1 + 0.20)m_b}{m_B} = \frac{1.2(0.15\,\text{kg})}{2.0\,\text{kg}} = 0.090 = 9.0\%$$

Assess: Because of the relative masses, this seems reasonable.

P9.71. Prepare: The ^{14}C atom undergoes an "explosion" and decays into a nucleus, an electron, and a neutrino. Due to the lack of external forces acting on the carbon atom, momentum is conserved in the process of "explosion" or decay. The initial momentum of the ^{14}C atom is zero. We will assume that explosion causes the electron to move along the x axis and the neutrino along the y axis. Also we note that the mass of the nucleus is essentially the same as the ^{14}C atom because the masses of the electron and the neutrino are very small. From the given data, we will first calculate the momentum of the nucleus and then divide this quantity by neutron's mass to obtain its speed.

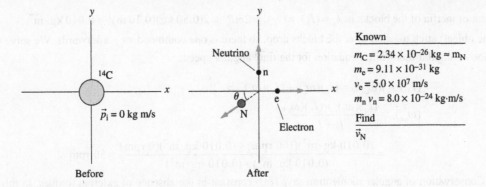

Before   After

Known

$m_C = 2.34 \times 10^{-26}$ kg $\approx m_N$

$m_e = 9.11 \times 10^{-31}$ kg

$v_e = 5.0 \times 10^7$ m/s

$m_n v_n = 8.0 \times 10^{-24}$ kg·m/s

Find

$\vec{v}_N$

Solve: The conservation of momentum equation $\vec{p}_f = \vec{p}_i = 0$ kg·m/s is

$$\vec{p}_e + \vec{p}_n + \vec{p}_N = 0 \text{ N} \Rightarrow \vec{p}_N = -(\vec{p}_e + \vec{p}_n) = -m_e \vec{v}_e - m_n \vec{v}_n$$

$$= -[(9.11 \times 10^{-31} \text{ kg})(5 \times 10^7 \text{ m/s}), \text{ along} + x] - [(8.0 \times 10^{-24} \text{ kg·m/s}), \text{ along} + y]$$

$$= -(45.55 \times 10^{-24} \text{ kg·m/s}, \text{ along} + x) - (8.0 \times 10^{-24} \text{ kg·m/s}, \text{ along} + y)$$

$$\Rightarrow p_N = m_N v_N = \sqrt{(45.55 \times 10^{-24})^2 + (8.0 \times 10^{-24})^2} \text{ kg·m/s} = 4.62 \times 10^{-23} \text{ kg·m/s}$$

$$\Rightarrow (2.34 \times 10^{-26} \text{ kg})v_N = 4.62 \times 10^{-23} \text{ kg·m/s} \Rightarrow v_N = 1.97 \times 10^3 \text{ m/s}$$

Assess: The nucleus speed is 4 orders of magnitude smaller than the electron's speed. This difference is primarily due to 4 orders of magnitude larger mass of the nucleus compared to the electron.

P9.73. Prepare: Model the puck as a particle.

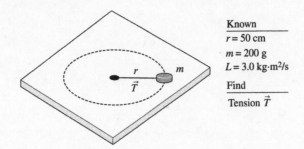

Known

$r = 50$ cm

$m = 200$ g

$L = 3.0$ kg·m²/s

Find

Tension $\vec{T}$

Solve: The angular momentum is

$$L = mvr = 3.0 \text{ kg·m}^2/\text{s} = (0.2 \text{ kg})v (0.5 \text{ m}) \Rightarrow v = 30 \text{ m/s}$$

The force that keeps the puck in circular motion is the tension $\vec{T}$ in the string. Thus,

$$T = \frac{mv^2}{r} = \frac{(0.2 \text{ kg})(30 \text{ m/s})^2}{0.5 \text{ m}} = 360 \text{ N}$$

Assess: A tension of 360 N in the string whose farthest end has a puck moving at approximately 60 mph is large, but in view of the puck's high speed the tension would be reasonable.

P9.75. Prepare: Model the turntable as a rigid disk rotating on frictionless bearings. As the blocks fall from above and stick on the turntable, the turntable slows down due to increased rotational inertia of the (turntable + blocks) system. Any torques between the turntable and the blocks are internal to the system, so angular momentum of the system is conserved.

$$(I_1)_f(\omega_1)_f + (I_2)_f(\omega_2)_f + \ldots = (I_1)_i(\omega_1)_i + (I_2)_i(\omega_2)_i + \ldots$$

The moment of inertia of the disk is $I_d = (I_d)_i = (I_d)_f = \frac{1}{2}mR^2 = \frac{1}{2}(2.0 \text{ kg})(0.10 \text{ m})^2 = 0.010$ kg·m².

The moment of inertia of the blocks is $I_b = (I_b)_i = (I_b)_f = 2mR^2 = 2(0.50 \text{ kg})(0.10 \text{ m})^2 = 0.010 \text{ kg} \cdot \text{m}^2$.

Solve: The objects stick together after the blocks drop, so there is one combined ω_{tot} afterwards. We solve the conservation of angular momentum equation for the final angular speed.

$$(I_d + I_b)(\omega_{tot})_f = (I_d)(\omega_d)_i + (I_b)(\omega_b)_i \Rightarrow$$

$$(\omega_{tot})_f = \frac{(I_d)(\omega_d)_i + (I_b)(\omega_b)_i}{I_d + I_b}$$

$$= \frac{(0.010 \text{ kg} \cdot \text{m}^2)(100 \text{ rpm}) + (0.010 \text{ kg} \cdot \text{m}^2)(0 \text{ rpm})}{(0.010 \text{ kg} \cdot \text{m}^2) + (0.010 \text{ kg} \cdot \text{m}^2)} = 50 \text{ rpm}$$

Assess: Conservation of angular momentum says $I\omega =$ constant in the absence of external torques. In this problem the moment of inertia goes up and therefore the angular frequency has to decrease, which is consistent with what we found.

ENERGY AND WORK

Q10.3. Reason: We must think of a process that increases an object's kinetic energy without increasing any potential energy. Consider pulling an object across a level floor with a constant force. The force does work on the object, which will increase the object's kinetic energy. Since the floor is level the gravitational potential energy does not change. The other form of potential energy possible is that stored in a spring, which is also zero here.

Assess: For there to be no potential energy change, the object in question must remain at the same height.

Q10.5. Reason: The system must convert kinetic energy directly to potential energy with no external force doing work. For gravitational potential energy we must change the height of the object. One simple example would be rolling a ball up a hill. The initial kinetic energy is converted to gravitational potential energy as the ball increases its height. The ball loses kinetic energy while it gains potential energy. Another example is rolling a ball into an uncompressed spring on level ground. As the ball compresses the spring, the system gains potential energy, while losing kinetic energy. Since there are no forces external to the systems in these examples, no work is done on the systems by the environment.

Assess: As long as no forces external to the system are applied, work done on a system is zero.

Q10.7. Reason: Here we need to convert potential energy to kinetic energy without any work done on the system. Consider dropping a ball from a height. The ball's gravitational energy is converted to the kinetic energy of the ball as it falls. Another example would be releasing a ball at the end of a compressed spring. The potential energy stored in the compressed spring is converted to the kinetic energy of the ball as the spring stretches to its equilibrium length. Since no external forces act on a system, the work on the system is zero.

Assess: Many examples in the problem section will involve just this type of conversion of potential energy to kinetic energy. If no forces from the environment act on a system, the work done on the system is zero.

Q10.11. Reason: The energies involved here are kinetic energy, gravitational potential energy, elastic potential energy, and thermal energy. For the system to be isolated, we must not have any work being done on the system and no heat being transferred into or out of the system. The ball's kinetic and elastic energy is changing, so we should consider it part of the system. Since its gravitational potential energy is changing, we need to also consider the earth as part of the system. Thermal energy will be generated in the ball and floor when the ball hits the floor, so we must consider both to be part of the system.

Assess: In order to have an isolated system no work can be done on the system and all forces must be internal.

Q10.13. Reason: (a) The work done is $W = Fd$. Both particles experience the same force, so the greater work is done on the particle that undergoes the greater displacement. Particle A, which is less massive than B, will have the greater acceleration and thus travel further during the 1 s interval. Thus more work is done on particle A. **(b)** Impulse is $F\Delta t$. Both particles experience the same force F for the same time interval $\Delta t = 1$ s. Thus the same impulse is delivered to both particles. **(c)** Both particles receive the same impulse, so the change in their momenta is the same, that is, $m_A(v_f)_A = m_B(v_f)_B$. But because $m_A < m_B$, it must be that $(v_f)_A > (v_f)_B$. This result can also be found from kinematics, as in part (a).

Assess: Work is the product of the force and the displacement, while impulse is the product of the force and the time during which the force acts.

Q10.15. Reason: Neglecting frictional losses, the work you do on the jack is converted into gravitational potential energy of the car as it is raised. The work you do is Fd, where F is the force you apply to the jack handle and d is the 20 cm distance you move the handle. This work goes into increasing the potential energy by an amount $mgh = wh$, where w is the car's weight and $h = 0.2$ cm is the change in the car's height. So $Fd = wh$ so that $F/w = h/d$.

Assess: Because the force F you can apply is so much less than the weight w of the car, h must be much less than d.

Q10.21. Reason: As you land, the force of the ground or pad does negative work on your body, transferring out the kinetic energy you have just before impact. This work is $-Fd$, where d is the distance over which your body stops. With the short stopping distance involved upon hitting the ground, the force F will be much greater than it is with the long stopping distance upon hitting the pad.

Assess: For a given amount of work, the force is large when the displacement is small.

Q10.23. Reason: When the coaster is at the top $U = mgy$ relative to the ground. That amount of energy equals the kinetic energy at the bottom. Halfway down (or up) the potential energy is half of what it was at the top, so the kinetic energy must also be half of what it is at the bottom. If v' is the speed at the halfway height, then

$$\frac{\frac{1}{2}m(v')^2}{\frac{1}{2}m(v)^2} = \frac{1}{2} \Rightarrow \left(\frac{v'}{v}\right)^2 = \frac{1}{2} \Rightarrow \frac{v'}{v} = \sqrt{\frac{1}{2}} \Rightarrow v' = \frac{\sqrt{2}}{2}(30 \text{ m/s}) = 21 \text{ m/s}$$

So the correct choice is C.

Assess: Even though the height is half the total, the speed is not half of 30 m/s.

Q10.25. Reason: Assuming the woman raises the weight at constant velocity, the force she exerts must equal the weight of the object. Since the mass of the object is 20 kg, its weight is about $w = mg = (20 \text{ kg})(9.80 \text{ m/s}^2) = 200 \text{ N}$. Since she lifts it 2 m, the work done is $W = Fd = (200 \text{ N})(2 \text{ m}) = 400 \text{ J}$. She does this work in 4 s, so the power she exerts is $P = W/\Delta t = (400 \text{ J})/(4 \text{ s}) = 100 \text{ W}$. The correct choice is A.

Assess: Power is defined as the rate of doing work. This seems like a reasonable amount of power for the woman to expend.

Problems

P10.1. Prepare: Equation 10.6 gives the work done by a force $\vec{F}$ on a particle. The work is defined as $W = Fd\cos(\theta)$, where d is the particle's displacement. Since there is a component of the lifting force in the direction of the displacement, we expect the work done in both parts of this problem to be nonzero. Assume you lift the book steadily, so that the force exerted on the book is constant.

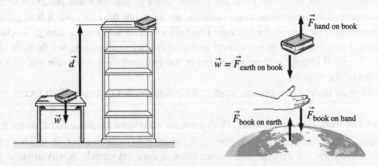

Solve: (a) Refer to the diagram. We are assuming the book does not accelerate, so the force you exert on the book is exactly equal to the force of gravity on the book. $\vec{F}_{\text{hand on book}} = -\vec{F}_{\text{gravity on book}}$

The total displacement of the book is $2.3 \text{ m} - 0.75 \text{ m} = 1.55 \text{ m}$ (keeping one extra significant figure for this intermediate result).

The work done by gravity is then

$$W_{\text{gravity on book}} = wd\cos(\theta) = (2.0 \text{ kg})(9.80 \text{ m/s}^2)(1.55 \text{ m})\cos(180°) = -30 \text{ J}$$

(b) The work done by hand is $W_{\text{hand on book}} = F_{\text{hand on book}}d\cos(\theta)$.

$$\Rightarrow W_{\text{hand on book}} = (2.0 \text{ kg})(9.80 \text{ m/s}^2)(1.55 \text{ m})\cos(0°) = +30 \text{ J}$$

Assess: Note that the only difference is in the sign of the answer. This is because the two forces are equal, but act in opposite directions. The work done by gravity is negative because gravity acts opposite to the displacement of the book. Your hand exerts a force in the same direction as the displacement, so it does positive work. We should expect the total work to be zero from Equation 10.4 since energy is conserved in this process. Referring to the results, we see that the work by your hand cancels the work done by gravity and the total work is zero as expected.

P10.5. Prepare: Equation 10.6 is the definition of work when the force and displacement are not parallel, as is the case in this problem.

Solve: (a) The boy is standing still in this case, so the displacement is zero. $W = Fd\cos(\theta) = (F)(0 \text{ m})\cos(\theta) = 0 \text{ J}$.

The work done on the boy by the string is exactly zero Joules.

(b) The displacement is non-zero in this case, so we expect the work done to be non-zero. Refer to the following figure. The angle between the force and the displacement is $180° - 30° = 150°$.

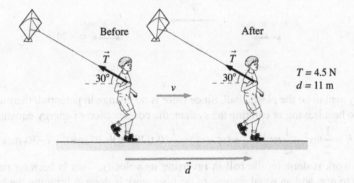

The work is

$$W = Fd\cos(\theta) = (4.5 \text{ N})(11 \text{ m})\cos(150°) = -43 \text{ J}$$

(c) The angle between the force and displacement in this case is $30°$ (Look at the figure and imagine the direction of the displacement vector is reversed). The work is

$$W = Fd\cos(\theta) = (4.5 \text{ N})(11 \text{ m})\cos(30°) = 43 \text{ J}$$

Assess: For there to be work done, the displacement must not be zero. If there is no displacement there is no work done. Note that the answers to parts **(b)** and **(c)** have opposite signs. This is because the displacement is exactly opposite in those cases for the same direction of the force.

P10.7. Prepare: The kinetic energy for any object moving of mass m with velocity v is given in Equation 10.8: $K = \dfrac{1}{2}mv^2$.

$v_B = 500 \text{ m/s}$
$m_B = 10 \text{ g}$

$v_{BB} = 10 \text{ m/s}$
$m_{BB} = 10 \text{ kg}$

Solve: For the bullet,

$$K_B = \frac{1}{2}m_B v_B^2 = \frac{1}{2}(0.010 \text{ kg})(500 \text{ m/s})^2 = 1.3 \text{ kJ}$$

For the bowling ball,

$$K_{BB} = \frac{1}{2} m_{BB} v_{BB}^2 = \frac{1}{2}(10 \text{ kg})(10 \text{ m/s})^2 = 0.50 \text{ kJ}$$

Thus, the bullet has the larger kinetic energy.

Assess: Kinetic energy depends not only on mass but also on the square of the velocity. The previous calculation shows this dependence. Although the mass of the bullet is 1000 times smaller than the mass of the bowling ball, its speed is 50 times larger, which leads to the bullet having over twice the kinetic energy of the bowling ball.

P10.11. Prepare: Use the definition of kinetic energy.
Solve: The man and the bullet have the same kinetic energy.

$$\tfrac{1}{2} m_m v_m^2 = \tfrac{1}{2} m_b v_b^2 \Rightarrow$$

$$v_m = \sqrt{\frac{m_b}{m_m}}\, v_b = \sqrt{\frac{8.0 \text{ g}}{80 \text{ kg}}}(400 \text{ m/s}) = 4.0 \text{ m/s}$$

Assess: We expected the man to need much less speed than the bullet to have the same kinetic energy.

P10.13. Prepare: Use the law of conservation of energy, Equation 10.4, to find the work done on the particle. We will assume there is no change in thermal energy of the ball.

$$v_f = 30 \text{ m/s} \qquad \vec{F} \qquad v_i = 30 \text{ m/s}$$
$$\longrightarrow \qquad \longleftarrow \bullet \; m = 20 \text{ g}$$

After   Before

Solve: Consider the system to be the plastic ball. Since there is no change in potential, thermal or chemical energy of the ball, and there is no heat leaving or entering the system, the conservation of energy equation becomes

$$W = \Delta K = \frac{1}{2} m v_f^2 - \frac{1}{2} m v_i^2 = \frac{1}{2} m (v_f^2 - v_i^2) = \frac{1}{2}(0.020 \text{ kg})[(30 \text{ m/s})^2 - (-30 \text{ m/s})^2] = 0 \text{ J}$$

Assess: Note that no work is done on the ball in reversing its velocity. This is because negative work is done in slowing the ball down to rest, and an equal amount of positive work is done in bringing the ball to the original speed but in the opposite direction.

P10.15. Prepare: Energy is stored in the flywheel by virtue of the motion of the particles and is given by Equation 10.9. In this equation, units for rotational velocity must be rad/s.
Solve: Using Equation 10.9,

$$K_{rot} = \frac{1}{2} I \omega^2, \text{ so } I = \frac{2K_{rot}}{\omega^2}$$

We need to convert ω to proper units, radians/s. Since $\omega = 20\,000$ rev/min and there are 2π rad/rev and 60 s/min,

$$\omega = \left(20\,000 \frac{\text{rev}}{\text{min}}\right)\left(\frac{1 \text{ min}}{60 \text{ s}}\right)\left(\frac{2\pi \text{ rad}}{\text{rev}}\right)$$

So.

$$I = \frac{(2)(4.0 \times 10^6 \text{ J})}{\left[\left(20\,000 \dfrac{\text{rev}}{\text{min}}\right)\left(\dfrac{1 \text{ min}}{60 \text{ s}}\right)\left(\dfrac{2\pi \text{ rad}}{\text{rev}}\right)\right]^2} = 1.8 \text{ kg} \cdot \text{m}^2$$

Assess: The flywheel can store this large amount of energy even though it has a low moment of inertia because of its high rate of rotation.

P10.21. Prepare: Assume an ideal spring that obeys Hooke's law. Equation 10.15 gives the energy stored in a spring. The elastic potential energy of a spring is defined as $U_s = \frac{1}{2}kx^2$, where x is the magnitude of the stretching or compression relative to the unstretched or uncompressed length. $\Delta U_s = 0$ when the spring is at its equilibrium length and $x = 0$.

Solve: We have $U_s = 200$ J and $k = 1000$ N/m. Solving for x:

$$x = \sqrt{2U_s/k} = \sqrt{2(200 \text{ J})/(1000 \text{ N/m})} = 0.63 \text{ m}$$

Assess: In the equation for the elastic potential energy stored in a spring, it is always the distance of the stretching of compression relative to the *unstretched* or *equilibrium* length.

P10.23. Prepare: We will assume the knee extensor tendon behaves according to Hooke's Law and stretches in a straight line. The elastic energy stored in a spring is given by Equation 10.15, $U_s = \frac{1}{2}kx^2$.

Solve: For athletes,

$$U_{s,\text{athlete}} = \frac{1}{2}kx^2 = \frac{1}{2}(33\,000 \text{ N/m})(0.041 \text{ m})^2 = 27.7 \text{ J}$$

For non-athletes,

$$U_{s,\text{non-athlete}} = \frac{1}{2}kx^2 = \frac{1}{2}(33\,000 \text{ N/m})(0.033 \text{ m})^2 = 18.0 \text{ J}$$

The difference in energy stored between athletes and non-athletes is therefore 9.7 J.

Assess: Notice the energy stored by athletes is over 1.5 times the energy stored by non-athletes.

P10.27. Reason: The force of gravity and the force of friction are doing work on the child. Since the child slides at a constant speed, the net work (which is the change in kinetic energy) is zero. This allows us to write:

$$W_g + W_f = \Delta K = 0 \text{ or } W_f = -W_g$$

Knowing how the work done by gravity is related to the change in gravitational potential energy and how the work done by friction is related to the force of friction, we can determine the force of friction.

Solve: Writing expressions for the work done by friction obtain

$$W_f = -W_g = -(-\Delta U_g) = \Delta U_g = Mgh \text{ and } W_f = F_f L\cos 180° = -F_f L$$

Combining these and solving for the force of friction obtain

$$F_f = -Mgh/L = -(25\,\text{kg})(9.8\,\text{m/s}^2)(3.0\,\text{m})/(7.0\,\text{m}) = -1.0 \times 10^2 \text{ N}$$

The minus reminds us that the force of friction opposes the motion of the object, so the magnitude is 100 N.

Assess: A 100 N force of friction for a child sliding down a playground slid is a reasonable number.

P10.29. Prepare: The only force acting on the ball during its trip is gravity. The sum of the kinetic and gravitational potential energy for the ball, considered as a particle, does not change during its motion. Use Equation 10.4. Note that at the top of its trajectory when the ball turns around, the velocity of the ball is zero. Assume there is no friction.

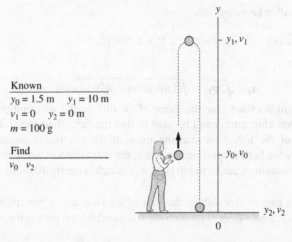

Known
$y_0 = 1.5$ m $y_1 = 10$ m
$v_1 = 0$ $y_2 = 0$ m
$m = 100$ g

Find
v_0 v_2

The figure shows the ball's before-and-after pictorial representation for the two situations described in parts (a) and (b).

Solve: Since energy is conserved, the quantity $K + U_g$ is the same during the entire trip. Thus, $K_f + U_{gf} = K_i + U_{gi}$.

(a) $\frac{1}{2}mv_1^2 + mgy_1 = \frac{1}{2}mv_0^2 + mgy_0 \Rightarrow v_0^2 = v_1^2 + 2g(y_1 - y_0)$

$\Rightarrow v_0^2 = (0 \text{ m/s})^2 + 2(9.80 \text{ m/s}^2)(10 \text{ m} - 1.5 \text{ m}) = 167 \text{ m}^2/\text{s}^2 \Rightarrow v_0 = 13 \text{ m/s}$

(b) $\frac{1}{2}mv_2^2 + mgy_2 = \frac{1}{2}mv_0^2 + mgy_0 \Rightarrow v_2^2 = v_0^2 + 2g(y_0 - y_2)$

$\Rightarrow v_2^2 = 167 \text{ m}^2/\text{s}^2 + 2(9.80 \text{ m/s}^2)(1.5 \text{ m} - 0 \text{ m}) \Rightarrow v_2 = 14 \text{ m/s}$

Assess: An increase in speed from 13 m/s to 14 m/s as the ball falls through a distance of 1.5 m is reasonable. Also, note that mass does not appear in the calculations that involve free fall since both gravitational potential energy and kinetic energy are proportional to mass. The mass cancels out in the equations.

P10.31. Prepare: The following figure shows a before-and-after pictorial representation of the rolling car. The car starts at rest from the top of the hill since it slips out of gear. Since we are ignoring friction, energy is conserved. The total energy of the car at the top of the hill is equal to total energy of the car at any other point.

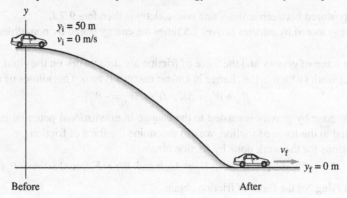

Solve: The energy conservation equation then becomes

$$K_i + (U_g)_i = K_f + (U_g)_f$$

or

$$\frac{1}{2}mv_i^2 + mgy_i = \frac{1}{2}mv_f^2 + mgy_f$$

The car starts from rest, so $v_i = 0$ m/s, which gives $K_i = 0$ J. Taking the bottom of the hill as the reference point for gravitational potential, $y_f = 0$ m and so $U_f = 0$ J.

The energy conservation equation becomes

$$\frac{1}{2}mv_f^2 = mgy_i$$

Canceling m and solving for v_f,

$$v_f = \sqrt{2gy_i} = \sqrt{(2)(9.80 \text{ m/s}^2)(50 \text{ m})} = 31 \text{ m/s}$$

Assess: Note that the problem does not give the shape of the hill, so the acceleration of the car is not necessarily constant. Constant acceleration kinematics can't be used to find the car's final speed. However, energy is conserved no matter what the shape of the hill. Note that the mass of the car is not needed. Since kinetic energy and gravitational potential energy are both proportional to mass, the mass cancels out in the equation. The final speed of the car, after traveling to the bottom of the 50 m hill is 31 m/s which is nearly 70 mi/h!

P10.35. Prepare: This is a case of free fall, so the sum of the kinetic and gravitational potential energy does not change as the rock is thrown. Assume there is no friction. The direction the rock is thrown is not known.

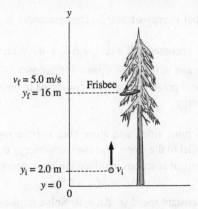

The coordinate system is put on the ground for this system, so that $y_f = 16$ m. The rock's final velocity v_f must be at least 5.0 m/s to dislodge the Frisbee.

Solve: The energy conservation equation for the rock $K_f + U_{gf} = K_i + U_{gi}$ is

$$\frac{1}{2}mv_f^2 + mgy_f = \frac{1}{2}mv_i^2 + mgy_i$$

This equation involves only the velocity magnitudes and not the angle at which the rock is to be thrown to dislodge the Frisbee. This equation is true for all angles that will take the rock to the Frisbee 16 m above the ground and moving with a speed of 5.0 m/s.
Using the previous equation we get

$$v_f^2 + 2gy_f = v_i^2 + 2gy_i \quad v_i = \sqrt{v_f^2 + 2g(y_f - y_i)} = \sqrt{(5.0 \text{ m/s})^2 + 2(9.80 \text{ m/s}^2)(16 \text{ m} - 2.0 \text{ m})} = 17 \text{ m/s}$$

Assess: Kinetic energy is defined as $K = \frac{1}{2}mv^2$ and is a scalar quantity. Scalar quantities do not have directional properties. Note also that the mass of the rock is not needed.

P10.37. Prepare: The thermal energy of the slide and the child's pants changes during the slide. If we consider the system to be the child and slide, total energy is conserved during the slide. The energy transformations during the slide are governed by the conservation of energy equation, Equation 10.4.

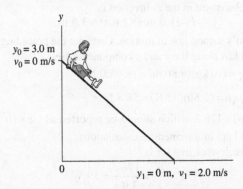

Solve: (a) The child's kinetic and gravitational potential energy will be changing during the slide. There is no heat entering or leaving the system, and no external work done on the child. There is a possible change in the thermal energy of the slide and seat of the child's pants. Use the ground as reference for calculating gravitational potential energy.

$$K_i = K_0 = \frac{1}{2}mv_0^2 = 0 \text{ J} \quad U_i = U_{g0} = mgy_0 = (20 \text{ kg})(9.80 \text{ m/s}^2)(3.0 \text{ m}) = 590 \text{ J}$$

$$W = 0 \text{ J} \quad K_f = K_1 = \frac{1}{2}mv_1^2 = \frac{1}{2}(20 \text{ kg})(2.0 \text{ m/s})^2 = 40 \text{ J} \quad U_f = U_{g1} = mgy_1 = 0 \text{ J}$$

At the top of the slide, the child has gravitational potential energy of 590 J. This energy is transformed partly into the kinetic energy of the child at the bottom of the slide. Note that the final kinetic energy of the child is only 40 J, much less than the initial gravitational potential energy of 590 J. The remainder is the change in thermal energy of the child's pants and the slide.

(b) The energy conservation equation becomes $\Delta K + \Delta U_g + \Delta E_{th} = 0$. With $\Delta U_g = -590$ J and $\Delta K = 40$ J, the change in the thermal energy of the slide and of the child's pants is then $590\text{ J} - 40\text{ J} = 550$ J.

Assess: Note that most of the gravitational potential energy is converted to thermal energy, and only a small amount is available to be converted to kinetic energy.

P10.39. Prepare: Call the system the tube, rider, and slope (but not the rope or rope puller). The snow is not frictionless. Assume the tow rope is parallel to the slope. Use the work-energy equation and $W = Fd$.

Solve: Since there are no springs or chemical reactions involved the work-energy equation is
$$W = \Delta K + \Delta U_g + \Delta E_{th}$$
We are told the towing takes place at a constant speed so $\Delta K = 0$. Solve for the change in thermal energy.
$$\Delta E_{th} = W - \Delta U_g = Fd - mg\Delta y = (340\text{ N})(120\text{ m}) - (80\text{ kg})(9.8\text{ m/s}^2)(30\text{ m}) = 17\text{ kJ}$$

Assess: W is the work done on the system by the rope.

P10.45. Prepare: We can use the definition of work, Equation 10.5, to calculate the work you do in pushing the block. The displacement is parallel to the force, so we can use $W = Fd$. Since the block is moving at a steady speed, the force you exert must be exactly equal and opposite to the force of friction.

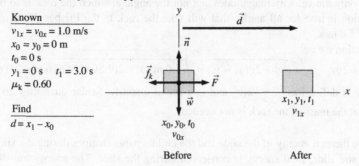

Solve: (a) The work done on the block is $W = Fd$ where d is the displacement. We will find the displacement using kinematic equations. The displacement in the x-direction is
$$d = (1.0\text{ m/s})(3.0\text{ s}) = 3.0\text{ m}$$
We will find the force using Newton's second law of motion. Consider the preceding diagram.
The equations for Newton's second law along the x and y components are
$$(F)_y = n - w = 0\text{ N} \Rightarrow n = w = mg = (10\text{ kg})(9.80\text{ m/s}^2) = 98.0\text{ N}$$
$$(F)_x = \vec{F} - \vec{f}_k = 0\text{ N} \Rightarrow F = f_k = \mu_k n = (0.60)(98\text{ N}) = 58.8\text{ N}$$
$$\Rightarrow W = Fd = (58.8\text{ N})(3.0\text{ m}) = 176\text{ J, which should be reported as }1.8 \times 10^2\text{ J to two significant figures.}$$
An extra significant figure has been kept in intermediate calculations.

(b) The power required to do this much work in 3.0 s is
$$P = \frac{W}{t} = \frac{176\text{ J}}{3.0\text{ s}} = 59\text{ W}$$

Assess: This seems like a reasonable amount of power to push a 10 kg block at 1.0 m/s. Note that this power is almost what a standard 60 W lightbulb requires!

P10.47. Prepare: The work done on the car while it is accelerating from rest to the final speed is the change in kinetic energy. Knowing the work done and the time to do this work we can determine the power associated with this work.

Solve: The change in kinetic energy of the car is
$$W = \Delta K = K_f - K_i = \frac{1}{2}mv_f^2 = \frac{1}{2}(1000\text{ kg})(30\text{ m/s})^2 = 4.5 \times 10^5\text{ J}$$
since the initial kinetic energy is zero.

The power associated with this work is

$$P = \frac{W}{\Delta t} = \frac{4.5 \times 10^5 \text{ J}}{10 \text{ s}} = 45 \text{ kW}$$

Assess: This is reasonable. In most cars only a small fraction of the work done by the engine goes into propelling the car.

P10.49. Prepare: Use the definition of power and solve for the time interval.
Solve: The change in energy is the change in gravitational potential energy.

$$P = \frac{\Delta E}{\Delta t} \Rightarrow \Delta t = \frac{\Delta E}{P} = \frac{mg\Delta y}{P} = \frac{(85 \text{ kg})(9.8 \text{ m/s}^2)(1100 \text{ m})}{450 \text{ W}} = 2036 \text{ s} \approx 34 \text{ min}$$

Assess: This is a reasonable length of time for a cylcist in a climbing stage in the Tour de France.

P10.51. Prepare: The two forces acting on the elevator are its weight and the force $\vec{F}$ due to the motor. Since the elevator is moving with constant velocity, the net force on the elevator is zero.
Solve: Since the net force on the elevator is zero, $\vec{F} + \vec{w} = \vec{0}$ N. So

$$F = -w = 2500 \text{ N}$$

The power due to this force acting on the elevator moving with constant velocity can be calculated using Equation 10.23.

$$P = Fv = (2500 \text{ N})(8.0 \text{ m/s}) = 2.0 \times 10^4 \text{ W}$$

Assess: One horsepower (hp) is 746 W, so the power of the motor is 26.8 hp. This is a reasonable amount of power to lift an elevator.

P10.55. Prepare: Assuming that the track offers no rolling friction, the sum of the skateboarder's kinetic and gravitational potential energy does not change during his rolling motion.

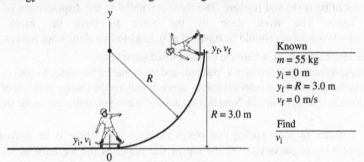

The vertical displacement of the skateboarder is equal to the radius of the track.
Solve: (a) The quantity $K \cdot U_g$ is the same at the upper edge of the quarter-pipe track as it was at the bottom. The energy conservation equation $K_f + U_{gf} = K_i + U_{gi}$ is

$$\frac{1}{2}mv_f^2 + mgy_f = \frac{1}{2}mv_i^2 + mgy_i \Rightarrow v_i^2 = v_f^2 + 2g(y_f - y_i),$$

$$v_i^2 = (0 \text{ m/s})^2 + 2(9.80 \text{ m/s}^2)(3.0 \text{ m} - 0 \text{ m}) = 58.8 \text{ m/s} \Rightarrow v_i = 7.7 \text{ m/s}$$

(b) If the skateboarder is in a low crouch, his height above ground at the beginning of the trip changes to 0.75 m. His height above ground at the top of the pipe remains the same since he is horizontal at that point. Following the same procedure as for part (a),

$$\frac{1}{2}mv_f^2 + mgy_f = \frac{1}{2}mv_i^2 + mgy_i \Rightarrow v_i^2 = v_f^2 + 2g(y_f - y_i)$$

$$v_i^2 = (0 \text{ m/s})^2 + 2(9.80 \text{ m/s}^2)(3.0 \text{ m} - 0.75 \text{ m}) = 44.1 \text{ m/s} \Rightarrow v_i = 6.6 \text{ m/s}$$

Assess: Note that we did not need to know the skateboarder's mass, as is the case with free-fall motion. Note that the shape of the track is irrelevant.

P10.59. Prepare: We will need to use Newton's laws here along with the definition of work (Equation 10.5). Assume you lift the box with constant speed.

Solve: (a) You lift the box with constant speed so the force you exert must equal the weight of the box. So $F = mg = (20 \text{ kg})(9.80 \text{ m/s}^2) = 196 \text{ N}$. The work done by this force is then $W = Fd = (196 \text{ N})(1.0 \text{ m}) = 196 \text{ J}$ which should be reported as 0.20 kJ to two significant figures

(b)

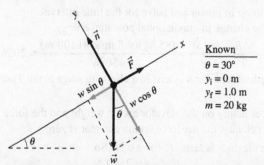

Known
$\theta = 30°$
$y_i = 0 \text{ m}$
$y_f = 1.0 \text{ m}$
$m = 20 \text{ kg}$

Refer to the preceding diagram. Since the box moves at constant speed, the force that is required to push the box up the ramp must exactly equal the component of the gravitational force along the slope.

$$F = mg\sin(\theta) = (20 \text{ kg})(9.80 \text{ m/s}^2)\sin(30°) = 98 \text{ N}$$

(c) Since the height of the ramp is 1.0 m and the angle of the ramp is 30°, the length of the ramp is the length of the hypotenuse in the diagram, which is

$$y = L\sin\theta \Rightarrow L = \frac{y}{\sin\theta} = \frac{1.0 \text{ m}}{\sin 30°} = 2.0 \text{ m}$$

(d) We will use the result of parts (b) and (c) here. The force is parallel to the displacement of the block, so we can use Equation 10.5 again. The work done by the force to push the block up the ramp is $W = Fd = (98 \text{ N})(2.0 \text{ m}) = 196 \text{ J}$ which should be reported as 0.20 kJ to two significant figures.

This is exactly the same result as part (a), where the block is lifted straight up.

Assess: We could have expected that the answers to parts (d) and (a) would be the same. In both cases the force we exert opposes gravity. We know that gravitational potential energy depends only on the change in height of an object, and not the exact path the object follows to change its height. Note that the answer doesn't even depend on the shape of the ramp.

P10.63. Prepare: Assume an ideal spring that obeys Hooke's law. There is no friction, and therefore the mechanical energy $K + U_s + U_g$ is conserved. At the top of the slope, as the ice cube is reversing direction, the velocity of the ice cube is 0 m/s.

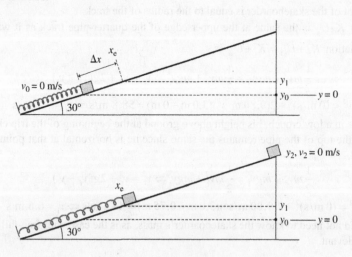

The figure shows a before-and-after pictorial representation. We have chosen to place the origin of the coordinate system at the position where the ice cube has compressed the spring 10.0 cm. That is, $y_0 = 0$.

Solve: The energy conservation equation $K_2 + U_{s2} + U_{g2} = K_0 + U_{s0} + U_{g0}$ is

$$\frac{1}{2}mv_2^2 + \frac{1}{2}k(x_e - x_e)^2 + mgy_2 = \frac{1}{2}mv_0^2 + \frac{1}{2}k(x - x_e)^2 + mgy_0$$

Using $v_2 = 0$ m/s, $y_0 = 0$ m, and $v_0 = 0$ m/s,

$$mgy_2 = \frac{1}{2}k(x - x_e)^2 \Rightarrow y_2 = \frac{k(x - x_e)^2}{2\,mg} = \frac{(25\ \text{N/m})(0.10\ \text{m})^2}{2(0.050\ \text{kg})(9.80\ \text{m/s}^2)} = 25.5\ \text{cm}$$

The distance traveled along the incline is $y_2 / \sin 30° = 51$ cm.

Assess: The net effect of the launch is to transform the potential energy stored in the spring into gravitational potential energy. The block has kinetic energy as it comes off the spring, but we did not need to know this energy to solve the problem since energy is conserved during the whole process.

P10.65. Prepare: This is a two-part problem. The slide is frictionless, so mechanical energy is conserved. We will calculate the final velocity of the people as they exit the slide and then use that result to calculate how far they travel from the exit before they hit the water.

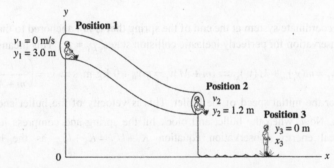

Solve: Refer to the diagram. Setting the reference for gravitational potential energy to be zero at the bottom of the slide, the energy conservation equation becomes

$$mgy_1 = \frac{1}{2}mv_1^2 \Rightarrow v_1 = \sqrt{2gy_1} = \sqrt{2(9.80\ \text{m/s}^2)(3.0\ \text{m})} = 7.67\ \text{m/s}$$

Note that this result does not depend on the person's mass. We keep an additional significant figure here for the second part of the calculation.

After they leave the slide, they are falling under the influence of gravity. Their initial velocity in the y direction is zero. The time it takes for them to fall to the water can be calculated with ordinary kinematics.

$$\Delta y = v_{0y}\Delta t - \frac{1}{2}a\Delta t^2, \text{ with } v_{0y} = 0\ \text{m/s gives}$$

$$\Delta y = -1.2\ \text{m} = -\frac{1}{2}(9.80\ \text{m/s}^2)\Delta t^2 \text{ or}$$

$$\Delta t = \sqrt{\frac{(2)(1.2\ \text{m})}{9.80\ \text{m/s}^2}} = 0.50\ \text{s}$$

Using $v_1 = 7.67$ s from the first part of the problem, we find

$$\Delta x = v_1\Delta t = (7.67\ \text{m/s})(0.50\ \text{s}) = 3.8\ \text{m}$$

The mass of the person was not necessary for this part of the calculation either.

Assess: Though this is a two-part problem mechanical energy is conserved throughout the whole process. However we could not use conservation of energy to solve the problem since we are not given the final velocity of the person before they hit the water, which is necessary for the conservation of energy equation. Note that it does not matter what the mass of the person is, they will always travel 3.8 m from the exit of the tube before hitting the water.

P10.67. Prepare: Assume an ideal spring that obeys Hooke's law. This is a two-part problem. The first part, when the bullet embeds itself in the block, is a perfectly inelastic collision. In a perfectly inelastic collision, the momentum is conserved while energy is not conserved. In the second part of the problem, when the bullet and block hit the spring, there is no friction. Since there is no friction after the bullet enters the block, the mechanical energy of the system (bullet + block + spring) is conserved during that part of the motion.

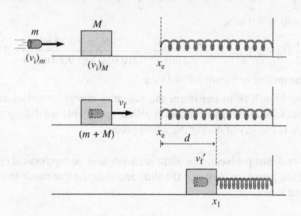

We place the origin of our coordinate system at the end of the spring that is not anchored to the wall.

Solve: (a) Momentum conservation for perfectly inelastic collision states $p_f = p_i$. This means

$$(m+M)v_f = m(v_i)_m + M(v_i)_M \Rightarrow (m+M)v_f = mv_B + 0 \text{ kg m/s} \Rightarrow v_f = \left(\frac{m}{m+M}\right)v_B$$

where we have used v_B for the initial speed of the bullet. This is velocity of the bullet and block after the bullet embeds itself in the block. Now, when the bullet and block hit the spring and compress it, mechanical energy is conserved. The mechanical energy conservation equation $K_1 + U_{s1} = K_e + U_{se}$ as the bullet-embedded block compresses the spring is:

$$\frac{1}{2}m(v_f')^2 + \frac{1}{2}k(x_1 - x_e)^2 = \frac{1}{2}(m+M)(v_f)^2 + \frac{1}{2}k(x_e - x_e)^2$$

$$0 \text{ J} + \frac{1}{2}kd^2 = \frac{1}{2}(m+M)\left(\frac{m}{m+M}\right)^2 v_B^2 + 0 \text{ J} \Rightarrow v_B = \sqrt{\frac{(m+M)kd^2}{m^2}}$$

(b) Using the preceding formula with $m = 5.0$ g, $M = 2.0$ kg, $k = 50$ N/m, and $d = 10$ cm,

$$v_B = \sqrt{\frac{(0.0050 \text{ kg} + 2.0 \text{ kg})(50 \text{ N/m})(0.10 \text{ m})^2}{(0.0050 \text{ kg})^2}} = 200 \text{ m/s}$$

which should be reported as 2.0×10^2 m/s to two significant figures.

(c) The fraction of energy lost is (initial energy − final energy)/(initial energy), which is

$$\frac{\frac{1}{2}mv_B^2 - \frac{1}{2}(m+M)v_f^2}{\frac{1}{2}mv_B^2} = 1 - \frac{m+M}{m}\left(\frac{v_f}{v_B}\right)^2 = 1 - \frac{m+M}{m}\left(\frac{m}{m+M}\right)^2$$

$$= 1 - \frac{m}{m+M} = 1 - \frac{0.0050 \text{ kg}}{(0.0050 \text{ kg} + 2.0 \text{ kg})} = 99.8\%$$

where we have kept an additional significant figure.

Assess: During the perfectly inelastic collision 99.8% of the bullet's energy is lost. The energy is transformed into the energy needed to deform the block and bullet and to the thermal energy of the bullet and block combination.

P10.69. Reason: If we consider the system of interest to be the two masses and the pulley, the only outside force doing work on the system is gravity. Since gravity is a conservative force, energy is conserved and we may write

$$\Delta K + \Delta U_g = \Delta E = 0 \quad \text{or} \quad \Delta K = -\Delta U_g.$$

Solve: Knowing $\Delta K = -\Delta U_g$, we may write

$$(M_A + M_B)v^2/2 = -(-M_B gh)$$

or

$$v = [2M_B gh/(M_A + M_B)]^{1/2} = \sqrt{[2(4.0 \text{ kg})(9.8 \text{ m/s}^2)(0.50 \text{ m})/(12 \text{ kg} + 4.0 \text{ kg})]} = 1.6 \text{ m/s}$$

Assess: This is a reasonable speed for this situation.

P10.73. Prepare: Initially the spring is hanging at its equilibrium position and you hang the fish on the spring, but don't release it. Let's agree to call this position 1 and to make this the point for zero gravitational and elastic potential energy. For the case where the fish is at position 1, we can write the following:

The position of the fish is $y_1 = 0$

The speed of the fish $v_1 = 0$

The kinetic energy of the fish $K_1 = 0$

The gravitational potential energy $U_{g_1} = 0$

The elastic potential energy of the spring $U_{S_1} = 0$

Total energy at this point is $E_1 = K_1 + U_{g_1} + U_{S_1}$

When the fish is released it falls to position 2, stretching the spring a maximum amount. For the case where the fish is at position 2, we can write the following:

The position of the fish is $y_2 = -h$

The speed of the fish $v_2 = 0$

The kinetic energy of the fish $K_2 = 0$

The gravitational potential energy $U_{g_2} = -mgh$

The elastic potential energy of the spring $U_{S_1} = kh^2/2$

Total energy at this point is $E_2 = K_2 + U_{g_2} + U_{S_2}$

Knowing that energy is conserved ($E_1 = E_2$), we can determine the maximum distance the fish falls.

Solve: Inserting values for the kinetic energy, gravitational potential energy and the elastic potential energy into

$$E_1 = E_2$$

obtain

$$0 = -mgh + kh^2/2$$

or

$$h = 2mg/k = 2(5.0 \text{ kg})(9.8 \text{ m/s}^2)/200 \text{ N/m} = 0.49 \text{ m}$$

Assess: This is a reasonable amount for the maximum stretch of the spring.

P10.77. Prepare: We will use the constant-acceleration kinematic equations and the definition of power in terms of work, Equation 10.22.

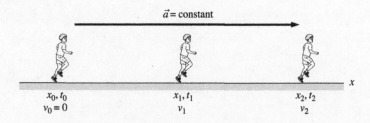

Solve: Refer to the diagram.

(a) We can find the acceleration from the kinematic equations and the horizontal force from Newton's second law. We have

$$x_2 = x_o + v_0(t_2 - t_0) + \frac{1}{2}a(t_2 - t_0)^2 \Rightarrow 50 \text{ m} = 0 \text{ m} + 0 \text{ m} + \frac{1}{2}a(7.0 \text{ s} - 0 \text{ s})^2 \Rightarrow a = 2.04 \text{ m/s}^2$$

$$\Rightarrow F = ma = (50 \text{ kg})(2.04 \text{ m/s}^2) = 102 \text{ N}$$

which should be reported as 1.0×10^2 N to two significant figures.

(b) We obtain the sprinter's power output by using $P = \frac{W}{\Delta t}$, where W is the work done by the sprinter. After $t = 2.0$ s, the sprinter has moved a distance of $d = \frac{1}{2}(2.04 \text{ m/s}^2)(2.0 \text{ s})^2 = 4.08$ m. The work done by the sprinter is then $W = Fd = (102 \text{ N})(4.08 \text{ m}) = 416$ J. His power output is then $P = \frac{W}{\Delta t} = \frac{416 \text{ J}}{2.0 \text{ s}} = 208$ W, or 210 W to two significant figures.

(c) During the final two seconds of his run, the distance he has moved is given by

$$d = x_2 - x_1 = v_1(t_2 - t_1) + \frac{1}{2}a(t_2 - t_1)^2 = v_1(2.0 \text{ s}) + \frac{1}{2}(2.04 \text{ m/s}^2)(2.0 \text{ s})^2$$

We need his velocity at 5.0 s after he starts running, which is given by

$$v_1 = v_0 + a\Delta t = 0 \text{ m/s} + (2.04 \text{ m/s}^2)(5.0 \text{ s}) = 10.2 \text{ m/s}$$

So

$$d = (10.2 \text{ m/s})(2.0 \text{ s}) + \frac{1}{2}(2.04 \text{ m/s}^2)(2.0 \text{ s})^2 = 24.5 \text{ m}$$

The work done by the sprinter is $W = Fd = (102 \text{ N})(24.5 \text{ m}) = 2.50$ kJ. His power output is

$$P = \frac{W}{\Delta t} = \frac{2.50 \text{ kJ}}{2.0 \text{ s}} = 1.2 \text{ kW}$$

Assess: Note the power output required for the last two seconds of the sprint is much larger than during the first two seconds. This is because the sprinter travels a much larger distance during the last two seconds of his trip because he has accelerated to a high velocity by that time. The force is the same during both time intervals.

USING ENERGY

Q11.3. Reason: Riding a bike, you do not change your height. As you run you constantly propel yourself off the ground. Extra energy is required for this.

Q11.5. Reason: We know from the previous chapter that kinetic energy increases as the square of the velocity. As velocity increases, kinetic energy increases. As the engine turns faster, the rate at which friction creates waste thermal energy also increases. Also, as the speed of the car increases the drag force on the car increases. All of this requires a faster rate of energy consumption of the energy available as chemical energy in the fuel.

Q11.9. Reason: Since the blocks are at the same temperature, the average kinetic energy of the atoms that make up the block is the same. However, since the 3 kg block contains more atoms, it contains more thermal energy. If the blocks are placed in contact, the average kinetic energy of the atoms in each will remain the same, since there is no source of a higher average kinetic energy. The total energy of the combined system is the sum of the energies of both.
Assess: Thermal energy is directly related to average kinetic energy of the atoms or molecules that make up the system.

Q11.11. Reason: The system's temperature could stay the same if there is work done on the system to exactly compensate for the heat lost.
Assess: Remember both ways of changing the thermal energy of a system.

Q11.17. Reason: We can obtain this situation if we have a process where the amount of work done by a system is equal to the heat that comes into the system. This can be true for a gas that is heated and allowed to expand freely and do work as it expands.

Q11.21. Reason: As you do work on the rubber band, W is positive. The heat you feel from the rubber band is leaving the rubber band, so Q is negative.
Assess: For work done on a system W is positive. For heat leaving a system Q is negative.

Q11.23. Reason: It is possible to convert work into thermal energy with 100% efficiency. An example would be pushing a block across a table with friction at constant velocity. All of the work done goes into increasing the thermal energy of the block and table. Entropy increases, as required by the second law.
Assess: The second law says no heat engine (which converts thermal energy into useful work) can be 100% efficient, but the example given is not a heat engine.

Q11.27. Reason: The second law of thermodynamics means that the entropy of *everything* will increase during any process; small pockets of order can pop up if larger pockets of disorder appear elsewhere. In order to cool the inside of the freezer, you have run a reverse heat engine and dump a lot of heat into the kitchen around the freezer. The net entropy change for everything involved (the ice, the freezer, the kitchen, etc.) is greater than zero.

Assess: The ice cubes aren't isolated, as specified in the second law.

Q11.29. Reason: The source for energy in a person is the chemical energy in the food. Since the person is walking at a constant speed her kinetic energy is not changing. The answer is A.

Q11.33. Reason: The first law of thermodynamics is violated by this device since it converts a lesser amount of thermal energy to a greater amount of work output. This device also violates the second law of thermodynamics since it states that it is not possible to make a heat engine that converts thermal energy to an equivalent amount of work, and this device turns thermal energy to twice the amount of work! The answer is C.

Q11.35. Reason: The efficiency is $e = \dfrac{Q_H - Q_C}{Q_H}$ and the maximum efficiency is $e_{max} = 1 - \dfrac{T_C}{T_H}$. Set these equal to each other and solve for $Q_C \cdot T_C = 20\,^\circ\text{C} = 293\,\text{K}$ and $T_H = 450\,^\circ\text{C} = 723\,\text{K}$.

$$\frac{Q_H - Q_C}{Q_H} = 1 - \frac{T_C}{T_H}$$

$$1 - \frac{Q_C}{Q_H} = 1 - \frac{T_C}{T_H} \quad \Rightarrow \quad Q_C = Q_H \frac{T_C}{T_H} = (100\,\text{MJ})\left(\frac{293\,\text{K}}{723\,\text{K}}\right) = 40\,\text{MJ}\ .$$

So the answer is D.
Assess: Power plants do heat up rivers like this.

Problems

P11.3. Prepare: Efficiency is given by Equation 11.2

$$e = \frac{\text{what you get}}{\text{what you had to pay}}$$

In this case the $4.0 \times 10^{-3}\,\text{W}$ of electrical energy is "what you get" as visible light, and the $1.2 \times 10^{-1}\,\text{W}$ of light energy is "what you had to pay."
Solve: The efficiency calculation gives

$$e = \frac{4.0 \times 10^{-3}\,\text{W}}{1.2 \times 10^{-1}\,\text{W}} = 0.033 = 3.3\%$$

Assess: Photovoltaic (PV) cells, also known as solar cells, are notoriously inefficient, and this is a typical value for traditional cells. However, advances in technology have been made and efficiencies of up to 20% are available; some researchers are aiming for 40% efficiency.

P11.5. Prepare: This is simply a unit conversion problem. Remember that a food Calorie is 1000 physics calories.

Solve: $1000\,\text{Cal} = 1000\,\text{kcal} = 1000000\,\text{cal}\left(\dfrac{4.19\,\text{J}}{1\,\text{cal}}\right) = 4.19 \times 10^6\,\text{J} \approx 4.2 \times 10^6\,\text{J}$

Assess: A 1000 Cal burger has a *lot* of joules of energy! A small burger without the cheese and bacon could have significantly less than this, but still over a million joules.

P11.7. Prepare: Various fuels and the corresponding energy in 1 g of each are listed in Table 11.1. Fats in foods such as "energy bars" have an energy content of 38 kJ per gram. Since we have 6 g of fat in our energy bar, we simply need to multiply 6 g by 38 kJ/g.
Solve:

$$6\,\text{g}\left(\frac{38\,\text{kJ}}{1\,\text{g}}\right) = 228\,\text{kJ} \approx 230\,\text{kJ}$$

$$230\,\text{kJ}\left(\frac{1000\,\text{J}}{1\,\text{kJ}}\right) = 230000\,\text{J}$$

$$228000\,\text{J}\left(\frac{1\text{cal}}{4.19\,\text{J}}\right) = 54000\,\text{cal}$$

$$54000\,\text{cal}\left(\frac{1\text{kcal}}{1000\,\text{cal}}\right) = 54\,\text{kcal} = 54\,\text{Cal}$$

Assess: Comparing with Table 11.2 shows that the fat in the energy bar does not provide as much energy as a fried egg; however, the energy bar may also contain carbohydrates in addition, and fat provides more energy per gram than carbohydrates do, so the total number of food calories in the energy bar might be 150.

A 68 kg person needs just over 2000 Cal for basic life processes, so they would need to eat about 15 energy bars per day if that is all they ate.

You may have learned in a health or nutrition class that 1 g of fat provides about 9 Cal of energy. Our calculations above bear this out: $54.4\,\text{Cal}/6\,\text{g} = 9.0\,\text{Cal/g}$.

P11.11. Prepare: Table 11.4 tells us that a 68 kg person (we'll assume this is your mass) needs to expend 480 J/s to pedal a bicycle at a speed of 15 km/h.

Table 11.1 helps us calculate the chemical energy stored in one gallon of gasoline (which has a mass of 3.2 kg).

$$E_{\text{chem}} = 3.2\,\text{kg}\left(\frac{1000\text{g}}{1\text{kg}}\right)\left(\frac{44\text{kJ}}{1\text{g}}\right)\left(\frac{1000\text{J}}{1\text{kJ}}\right) \approx 1.4 \times 10^8\,\text{J}$$

Solve: The time that the chemical energy will last at the rate of 480 J/s is

$$\Delta t = \frac{\Delta E_{\text{chem}}}{P} = \frac{1.4 \times 10^8\,\text{J}}{480\,\text{W}} = 2.93 \times 10^5\,\text{s} = 81\,\text{h}$$

And the distance that can be covered during this time at 15 km/h is
$$\Delta x = v\Delta t = (15\,\text{km/h})(81\,\text{h}) \approx 1200\,\text{km}$$

to two significant figures.

Assess: The driving distance from Dallas, Texas to Denver, Colorado is just over 1200 km. This seems far for one gallon of gasoline, but you are going much slower than a car would (which increases your efficiency by decreasing the drag) and you are taking a lot less mass. Also remember that a car's efficiency is probably less than 10% as shown in Integrated Example 11.15, while the efficiency of a human cycling is 20%–30%.

To put it in units of mpg to compare to your car, you would be able to cycle 760 miles with the energy supplied by that one gallon of gas, which is 30 times better than a car that gets 25 mpg.

P11.13. Prepare: In weightlifting, a barbell curl is an exercise in which the barbell is held down at arms' length against the thighs and then raised in semi-circular motion until the forearms touch the biceps. We'll assume that the weightlifter expends metabolic energy when he lifts the 30 kg bar, but not as he lowers it. We'll also assume 25% efficiency.

Table 11.2 tells us that a typical slice of pizza has a metabolic energy content of 300 Cal or 1260 kJ.

The weightlifter will use the 1260 kJ (at 25% efficiency) to lift the 30 kg bar, increasing its potential energy. In one repetition he'll increase the potential energy by $\Delta U_{\text{g}} = mg\Delta y = mg(0.060\,\text{m})$, and in n repetitions by nmg (0.060 m), where n is what we want to know.

Solve:
$$\text{(energy from pizza) (efficiency)} = nmg\Delta y$$

$$(1.26 \times 10^6\,\text{J})(0.25) = n(30\,\text{kg})(9.80\,\text{m/s}^2)(0.60\,\text{m})$$

Solving for n gives 1790 repetitions, which should be reported to two significant figures as 1800 repetitions.

Assess: That's a lot of curls! Exercising in this way to "burn off" an extra slice of pizza is almost impossible; people can't do 1800 reps in a row, and there isn't time before the next meal anyway. And n would be four times larger if the weightlifter were 100% efficient!

P11.15. Prepare: Use the temperature conversion formula from Equation 11.5.
Solve:

$$T(^{\circ}\text{C}) = T(\text{K}) - 273 = 4.2 - 273 = -269\,^{\circ}\text{C}$$

$$T(^\circ F) = \frac{9}{5} T(^\circ C) = 32^\circ = \frac{9}{5}(-269\,^\circ C) + 32^\circ = -452\,^\circ F$$

Assess: That sounds like a record cold temperature.

P11.21. Prepare: Equation 11.8 gives the thermal energy change $\Delta E_{th} = W + Q$, and Figure 11.15 helps us figure out the signs. The 300 J of heat energy transferred to the system will be a positive Q, and the ΔE_{th} is a positive 150 J.
Solve:

$$W = \Delta E_{th} - Q = 150\,J - 300\,J = -150\,J$$

Assess: Since W is negative, it means that the work is done by the system instead of on it.

P11.23. Prepare: The efficiency of an engine is given by Equation 11.9.
Solve: (a) The work done by the engine per cycle is

$$W_{out} = Q_H - Q_C = 55\,kJ - 40\,kJ = 15\,kJ$$

(b) During each cycle, the heat transferred into the engine is $Q_H = 55$ kJ, and the heat exhausted is $Q_C = 40$ kJ. The thermal efficiency of the heat engine is

$$e = 1 - \frac{Q_C}{Q_H} = 1 - \frac{40\,kJ}{55\,kJ} = 0.27$$

Assess: We could have also gotten the answer to part (b) from part (a), $e = W_{out}/Q_H = (15\,J)/(55\,J) = 0.27$.

P11.25. Prepare: Assume that the heat engine follows a closed cycle. The efficiency of an engine is given by Equation 11.9.
Solve: The engine's efficiency is

$$e = \frac{W_{out}}{Q_H} = \frac{W_{out}}{Q_C + W_{out}} = \frac{200\,J}{600\,J + 200\,J} = 0.25 = 25\%$$

Assess: This is a reasonable efficiency.

P11.29. Prepare: $T_C = 20\,^\circ C + 273 = 293\,K$. The maximum efficiency of a heat engine is $e_{max} = 1 - \dfrac{T_C}{T_H}$. Solve for T_H.
Solve:

$$T_H = \frac{T_C}{1 - e_{max}} = \frac{293\,K}{1 - 0.60} = 733\,K = 460\,^\circ C$$

Assess: The hot reservoir has to be really quite hot to achieve that efficiency.

P11.31. Prepare: The COP of a refrigerator is given by the equation before Equation 11.11.
Solve: The coefficient of performance of the refrigerator is

$$COP = \frac{Q_C}{W_{in}} = \frac{Q_H - W_{in}}{W_{in}} = \frac{50\,J - 20\,J}{20\,J} = 1.5$$

P11.33. Prepare: The COP of a refrigerator is given by the equation just before Equation 11.11.
Solve: (a) The heat extracted from the cold reservoir is calculated as follows:

$$COP = \frac{Q_C}{W_{in}} \Rightarrow 4.0 = \frac{Q_C}{50\,J} \Rightarrow Q_C = 200\,J$$

(b) The heat exhausted to the hot reservoir is

$$Q_H = Q_C + W_{in} = 200\,J + 50\,J = 250\,J$$

P11.37. Prepare: We'll show all the possibilities and then directly count the probability that all three balls will be in Box 1.

Solve:

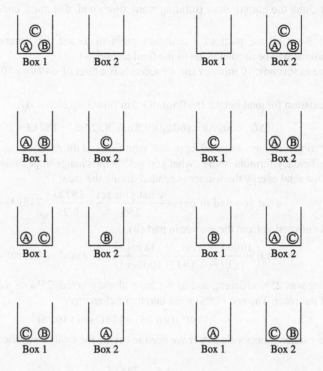

There are eight possible arrangements, and in only one of them are all three balls in Box 1. Therefore, the probability of that happening is $1/8 = 0.125 \approx 13\%$.

Assess: It is not very probable to have all three balls in Box 1 or all in Box 2; much more probable is to have two balls in one box and one ball in the other. This trend becomes more pronounced as the number of balls increases.

P11.39. Prepare: We see from Table 11.4 that running at 15 km/h requires about 1150 W of metabolic power for a 68 kg individual (hence that assumption for your mass as well). The speed of 15 km/h (a typical running speed) will also be used.

We will also estimate the efficiency at 25%, as is typical of walking, running, and climbing stairs.

Solve: Use $\text{energy} = \text{power} \times \text{time},$ modified to account for the efficiency of $e = 25\%$. Here the energy (in running) is what you had to "pay," and the power × time is what you get. So $\text{energy} \times e = \text{power} \times \text{time}$.

A couple of simple pre-conversions will make the following clearer and cleaner. $1150\,\text{W} = 1150\,\text{J/s}$ and $1\,\text{h} = 3600\,\text{s}$.

We also compute the time required for the marathon: $\Delta t = \dfrac{\Delta x}{v} = \dfrac{42.2\,\text{km}}{15\,\text{km/h}} = 2.813\,\text{h} = 10128\,\text{s}$.

$$E = P \times \Delta t = (1150\,\text{J/s})(10128\,\text{s}) = 1.165 \times 10^7\,\text{J}$$

Now we follow Example 11.5 to find the number of flights of stairs. At 25% efficiency, the amount of chemical energy transformed into increased potential energy is

$$\Delta U_g = (0.25)(1.165 \times 10^7\,\text{J}) = 2.912 \times 10^6\,\text{J}$$

Because $\Delta U_g = mg\Delta y$, the height gained is

$$\Delta y = \frac{\Delta U_g}{mg} = \frac{2.912 \times 10^6\,\text{J}}{(68\,\text{kg})(9.8\,\text{m/s}^2)} = 4\,370\,\text{m}$$

With each flight of stairs having a height of 2.7 m, the number of flights climbed is

$$\frac{4\,379\,\text{m}}{2.7\,\text{m}} = 1600\,\text{flights}$$

Assess: She could climb the Empire State Building many times with that much energy.

P11.43. Prepare: Read all the parts of a multipart problem to see the general direction of thought and what preliminary calculations will be needed to get to the final answer.

We convert the time to seconds: 10 minutes and 49 seconds is a total of $649s \approx 650s$.

Solve:

(a) Do a quick calculation for total height: $(86 \text{ floors})(3.7 \text{ m/floor}) = 320 \text{ m} = \Delta y$.

$$\Delta U_g = mg\Delta y = (60 \text{ kg})(9.8 \text{ m/s}^2)(320 \text{ m}) = 187 \text{ kJ} \approx 190 \text{ kJ}$$

This is true for the winner and any other 60 kg person who finished the race.

(b) We'll use the efficiency formula where "what you get" is the change in potential energy of the person, and "what you had to pay" is the total energy the winner expended during the race.

$$\text{what you had to pay} = \frac{\text{what you get}}{25\%} = \frac{187 \text{ kJ}}{0.25} = 748 \text{ kJ} \approx 750 \text{ kJ}$$

(c) This is just unit conversion from the answer in part **(b)**.

$$748 \text{ kJ}\left(\frac{1000 \text{ J}}{1 \text{ kJ}}\right)\left(\frac{1 \text{ cal}}{4.19 \text{ J}}\right)\left(\frac{1 \text{ kcal}}{1000 \text{ cal}}\right) = 179 \text{ kcal} \approx 180 \text{ kcal} = 180 \text{ Cal}$$

(d) Since the winner was 25% efficient, and as we have already noted, 25% of the energy went into increasing the potential energy of the racer. The rest (75%) went into thermal energy.

$$(179 \text{ Cal})(0.75) = 134 \text{ Cal} \approx 130 \text{ Cal}$$

(e) To compute the winner's metabolic power we need to divide the total metabolic energy "burned up" by the time it took.

$$P = \frac{\Delta E}{\Delta t} = \frac{748 \text{ kJ}}{650 \text{ s}} \approx 1200 \text{ W}$$

Assess: The winner of this race obviously made a good effort. In fact, the result of 1200 W is about one and a half horsepower.

P11.45. Prepare: In part **(a)** the swimmer's metabolic power is 800 W but the efficiency is only 25%, so the power that is used for forward propulsion is $0.25 \times 800 \text{ W} = 200 \text{ W}$.

The time it takes to swim 50 m is given as 22 s.

In part **(b)** we divide the 4400 J—half for the arms, and half for the legs. We further divide the arms' half by 30 to get the energy expenditure per arm stroke.

In part **(c)** the force will vary over the stroke, but we are only asked to determine the average force. The distance the hand moves through the water in a stroke (half a circle) is $\pi r = (0.90 \text{ m})\pi \approx 2.83 \text{ m}$.

The work done is the force multiplied by the distance, and the work done is the energy expended.

Solve: (a) $E = P\Delta t = (200 \text{ W})(22 \text{ s}) = 4400 \text{ J}$

(b) 2200 J/30 = 73 J per arm stroke.

(c) $\text{force} = \dfrac{\text{energy}}{\text{distance}} = \dfrac{73 \text{ J}}{2.83 \text{ m}} = 25.9 \text{ N}$

which should be reported to two significant figures as 26 N.

Assess: Swimming is strenuous, so 4400 J is not too much in 22 s.

73 J is about the energy needed to lift a 7.4 kg object vertically 1 m (the change in potential energy would be 73 J). This seems about as hard (energy intensive) as a swimming arm stroke.

Think about holding a 2.7 kg object at rest in your hand. The force you exert on it is about 26 N, which seems comparable to the force on the water in a swimming stroke.

P11.51. Prepare: We use ratios to solve the problem.

Solve:

(a) $(68 \text{ kg})\left(\dfrac{70000 \text{ Cal}}{5000 \text{ kg}}\right) = 952 \text{ Cal} \approx 950 \text{ Cal}$

(b) $P = \dfrac{E}{\Delta t} = \dfrac{952\,\text{Cal}}{1\,\text{d}}\left(\dfrac{1\,\text{d}}{24\,\text{h}}\right)\left(\dfrac{1\,\text{h}}{3600\,\text{s}}\right)\left(\dfrac{1000\,\text{cal}}{1\,\text{Cal}}\right)\left(\dfrac{4.19\,\text{J}}{1\,\text{cal}}\right) = 46\,\text{W}$

This is 54 W less than, or just under half the 100 W stated in the chapter.

Assess: The 46 W is less than a human, as we expect.

P11.53. Prepare: The theoretical maximum efficiency of a heat engine is given in terms of the hot reservoir temperature and the cold reservoir temperature by Equation 11.10.

$$e_{\text{max}} = 1 - \dfrac{T_C}{T_H}$$

Here we are given $T_H = 400\,\text{K}$ and $e = 0.20$ and asked to find the maximum possible T_C. This maximum occurs when $e = e_{\text{max}}$.

Solve Equation 11.10 for T_C:

$$\dfrac{(T_C)_{\text{max}}}{T_H} = 1 - e$$

$$(T_C)_{\text{max}} = T_H(1 - e)$$

Solve: With $T_H = 400\,\text{K}$ and $e = 0.20$,

$$(T_C)_{\text{max}} = T_H(1 - e) = (400\,\text{K})(1 - 0.20) = 320\,\text{K}$$

Assess: This answer is 80 K below T_H. (We'd be sure we had done something wrong if it had come out above T_H.)

P11.55. Prepare: Efficiency is given by Equation 11.9.

Solve: If $Q_C = \frac{2}{3}Q_H$, then $W_{\text{out}} = Q_H - Q_C = \frac{1}{3}Q_C$. Thus the efficiency is

$$e = \dfrac{W_{\text{out}}}{Q_H} = \dfrac{\frac{1}{3}Q_H}{Q_H} = \dfrac{1}{3}$$

The maxium efficiency of a heat engine is $e_{\text{max}} = 1 - T_C/T_H$. Thus

$$1 - \dfrac{T_C}{T_H} = \dfrac{1}{3} \quad \Rightarrow \quad \dfrac{T_C}{T_H} = \dfrac{2}{3}$$

P11.59. Prepare: The efficiency is what you get (the useful work output in this case) divided by what you had to pay (the thermal energy input in this case).

$$e = \dfrac{\text{what you get}}{\text{what you had to pay}} = \dfrac{W_{\text{out}}}{Q_H} = \dfrac{50\,\text{J}}{100\,\text{J}} = \dfrac{1}{2}$$

Equation 11.10 also says about efficiency that

$$e_{\text{max}} = 1 - \dfrac{T_C}{T_H}$$

The maximum efficiency will occur when the ratio of the temperatures (in kelvin) of the hot and cold reservoirs is a minimum. This minimum ratio is what we want to know.

Solve: Solve Equation 11.4 for $\frac{T_C}{T_H}$.

$$\left(\dfrac{T_C}{T_H}\right)_{\text{min}} = 1 - e_{\text{max}} = 1 - \dfrac{1}{2} = \dfrac{1}{2}$$

Assess: So 1/2 is the smallest possible ratio of the two temperatures in kelvin. This is certainly doable (say $T_C = 250\,\text{K}$ and $T_H = 500\,\text{K}$, although both temperatures won't be in a normal everyday weather-like range), so this engine is possible.

P11.61. Prepare: The power plant is to be treated as a heat engine.

Solve: (a) Every hour 300 metric tons or 3×10^5 kg of coal is burnt. The volume of coal is

$$\frac{3.00 \times 10^5 \, \text{kg}}{1 \, \text{hour}} \times \frac{1.5 \, \text{m}^3}{1000 \, \text{kg}} \times 24 \, \text{hour} = 1.08 \times 10^4 \, \text{m}^3$$

The height of the room will be 110 m.

(b) The thermal efficiency of the power plant is

$$e = \frac{W_{\text{out}}}{Q_{\text{H}}} = \frac{2.7 \times 10^{12} \, \text{J}}{(3.00 \times 10^5 \, \text{kg}) \times \dfrac{28 \times 10^6 \, \text{J}}{\text{kg}}} = 32\%$$

Assess: An efficiency of 32% is typical of power plants.

P11.65. Prepare: Say the outside temperature is $95\,^\circ\text{F} = 35\,^\circ\text{C} = 308 \, \text{K}$, and the desired indoor temperature is $75\,^\circ\text{F} = 24\,^\circ\text{C} = 297 \, \text{K}$.

Solve:

$$\text{COP}_{\text{max}} = \frac{T_{\text{C}}}{T_{\text{H}} - T_{\text{C}}} = \frac{297 \, \text{K}}{308 \, \text{K} - 297 \, \text{K}} = 27$$

$$\text{SEER}_{\text{max}} = 3.4 \times \text{COP}_{\text{max}} = (3.4)(27) = 92$$

Assess: Your answer may vary if you assumed different inside and outside temperatures.

CONSERVATION LAWS

PptII.5. Reason: The rubber bands slow down the rider over a longer period of time so the force is reduced. $F_{net}\Delta t = \Delta p$ shows this. The answer is D.

Assess: This is the same reason it doesn't hurt to land on a trampoline from a large height.

PptII.9. Reason: Since $F = kx$ reducing the spring constant will reduce the force on the rider. And since $\frac{1}{2}kx^2 = mgh$ reducing the spring constant will also reduce the final height (all else remaining equal). So the answer is A.

Assess: It makes sense that a smaller force would produce a lower jump height.

PptII.13. Reason: Use conservation of momentum, with the initial momentum of the system zero.

$$m_s v_s = -m_w v_w \quad \Rightarrow \quad v_s = \frac{-m_w v_s}{m_s} = \frac{-(0.30\,\text{kg})(10\,\text{m/s})}{4.0\,\text{kg}} = -0.75\,\text{m/s}$$

The squid's speed is thus 0.75 m/s, so the answer is D.

Assess: Because the water's mass is less than the squid's, we expect the squid's speed to be less than the water's.

PptII.15. Reason: Compute the average force from

$$F_{net} = \frac{\Delta p}{\Delta t} = \frac{m\Delta v}{\Delta t} = \frac{(0.30\,\text{kg})(10\,\text{m/s})}{0.10\,\text{s}} = 30\,\text{N}$$

So the answer is B.

Assess: This force is not very big.

PptII.19. Reason: See if the kinetic energy is the same before and after.

$$K_i = \frac{1}{2}m_c(v_c)_i^2 = \frac{1}{2}(0.30\,\text{kg})(40\,\text{m/s})^2 = 240\,\text{J}$$

$$K_f = \frac{1}{2}m_c(v_c)_f^2 + \frac{1}{2}m_b(v_b)_f^2 = \frac{1}{2}(0.30\,\text{kg})(30.3\,\text{m/s})^2 + \frac{1}{2}(0.046\,\text{kg})(63\,\text{m/s})^2 = 229\,\text{J}$$

Kinetic energy is not conserved, so the answer is B.

Assess: We would be worried if $K_f > K_i$.

PptII.21. Reason:

$$P = \frac{\Delta E}{\Delta t} = \frac{\frac{1}{2}mv^2}{\Delta t} = \frac{\frac{1}{2}(80\,\text{kg})(11\,\text{m/s})^2}{4.1\,\text{s}} = 1180\,\text{W} \approx 1200\,\text{W}$$

The answer is D.

Assess: It is hard to select between the choices based on ballpark expectations because they are all in a reasonable range.

PptII.25. Reason:

(a) Use conservation of momentum.

$$m_L(v_L)_i = (m_L + m_s)v_f \quad \Rightarrow \quad v_f = \frac{m_L(v_L)_i}{m_L + m_s} = \frac{(30\text{ kg})(4.0\text{ m/s})}{40\text{ kg}} = 3.0\text{ m/s}$$

(b)

$$F = \frac{\Delta p}{\Delta t} = \frac{(10\text{ kg})(3.0\text{ m/s})}{0.25\text{ s}} = 120\text{ N}$$

(c) Use $K = \frac{1}{2}mv^2$.

$$K_i = \frac{1}{2}(30\text{ kg})(4.0\text{ m/s})^2 = 240\text{ J} \qquad K_f = \frac{1}{2}(40\text{ kg})(3.0\text{ m/s})^2 = 180\text{ J}$$

$$\Delta K = K_f - K_i = 180\text{ K} - 240\text{ J} = -60\text{ J}$$

This energy was dissipated as thermal energy.

Assess: In part **(c)** it seems reasonable to "lose" a quarter of the energy.

THERMAL PROPERTIES OF MATTER

Q12.1. Reason: The mass of a mole of a substance in grams equals the atomic or molecular mass of the substance. Since neon has an atomic mass of 20, a mole of neon has a mass of 20 g. Since N_2 has a molecular mass of 28, a mole of N_2 has a mass of 28 g. Thus a mole of N_2 has more mass than a mole of neon.

Assess: Even though nitrogen *atoms* are lighter than neon atoms, nitrogen molecules are more massive, so a mole of nitrogen has more mass than a mole of neon.

Q12.3. Reason: Since there is almost pure helium in a helium balloon and almost no helium in the outside air, helium tends to diffuse out of the balloon. Similarly, with almost no oxygen or nitrogen in the balloon initially and high concentrations of oxygen and nitrogen in the air, these molecules tend to diffuse into the balloon. However, since helium atoms travel about three times faster than oxygen or nitrogen molecules and since helium atoms are smaller, they diffuse much faster so gas leaves the balloon faster than it enters. An air-filled balloon has the same particles inside as out and so the stated effect does not contribute to the deflation of such balloons. Instead, there is a weaker effect which is also at work for helium balloons: Higher pressure inside the balloon than outside makes the interior air molecules diffuse faster.

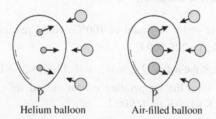

Helium balloon   Air-filled balloon

Assess: Because of the advantages helium atoms have in diffusion, helium balloons deflate faster than air-filled ones.

Q12.9. Reason: Since the bottles have the same temperature, the average kinetic energies of the atoms that are in each bottle is the same. Since both bottles have the same amount of atoms, the total thermal energy must be the same also.
Assess: Thermal energy in a gas is directly proportional to the number of atoms in the gas and its temperature.

Q12.11. Reason: Equation 12.14 applies. The number of molecules in the gas is constant since the container is sealed. Equation 12.14 can be written as $p = Nk_B(T/V)$.

(a) If the volume is doubled and the temperature tripled, the pressure increases by a factor of 3/2.
(b) If the volume is halved and the temperature tripled, the pressure increases by a factor of six.
Assess: This makes sense. Increasing the temperature increases the pressure in a gas as does decreasing the volume of the container.

Q12.13. Reason: If the work done is equals the area under the pV graph, then there is no work done if V is constant, because the graph would be a vertical straight line encompassing no area.

Stated another way, Equation 12.19 says $W_{gas} = p\Delta V$ and if V is constant then $\Delta V = 0$ and $W_{gas} = 0$.

Assess: As the chapter says, "in order for a gas to do work, the volume must change."

Q12.15. Reason: Thermal expansion will make lid expand. The glass jar will also expand, but the coefficient of expansion of the glass is smaller than the coefficient for the metal, so the lid expands more.

Assess: If you have never tried this trick, you should—and think of physics as you do.

Q12.21. Reason: The chocolate starts at some temperature. As it's heated the temperature rises to a point where the chocolate changes phase. Assuming the chocolate started as a solid, the chocolate melts during the second portion of the graph. During melting, the temperature is constant. After all the chocolate melts, the temperature rises as the liquid chocolate is heated.

Assess: Compare to Figure 12.22.

Q12.25. Reason: The thermal conductivity of copper is greater than that of stainless steel. The pot needs to distribute energy evenly over the bottom to cook the food evenly, so it needs to have high conductivity; the handle could have much lower thermal conductivity so it won't get so hot and burn hands.

Assess: Sometimes handles are made of even less conductive materials.

Q12.27. Reason: From the equation defining gauge pressure, $p = p_g + 1$ atm. The absolute pressure is higher inside the tire. The correct choice is C.

Assess: Gauge pressure is the pressure *above* atmospheric pressure.

Q12.31. Reason: In an adiabatic compression, the temperature rises. From the ideal gas law, we can say: $P = nRT/V$. If the temperature were constant, halving the volume would double the pressure. But since the temperature rises and temperature is in the numerator of the pressure formula, the pressure will more than double. The answer is D.

Assess: It makes sense that an adiabatic compression would raise the temperature more than an isothermal compression such as we saw in Q12.29. This is because adiabatic processes are generally processes which occur rapidly so that heat does not have time to escape. After the heat starts to escape, the gas particles have less energy and exert less pressure on the walls of the container.

Q12.35. Reason: To condense the steam to liquid at 100°C would require extracting an amount of energy of $Q_{cond} = ML_v = (0.1 \, \text{kg})(22.6 \times 10^5 \, \text{J/kg}) = 226\,000 \, \text{J}$.

The amount of energy required to melt the ice is $Q_{melt} = ML_f = (0.1 \, \text{kg})(3.33 \times 10^5 \, \text{J/kg}) = 33300 \, \text{J}$.

The amount of energy required to raise the temperature of the melted ice (now a liquid) from 0°C to 100°C is $Q_{\Delta T} = Mc\Delta T = (0.1 \, \text{kg})(4190 \, \text{J/kg} \cdot \text{K})(100 \, \text{K}) = 41900 \, \text{J}$.

The steam has more than enough energy to melt the ice and raise its temperature by 100°C. Some of the steam would not even need to condense to a liquid; it would stay as steam.

The correct answer is E.

Assess: The key to this question is the large difference between L_v and L_f for water. Had the difference been so large the other way the answer would have been A.

Q12.37. Reason: We'll first bring the ice up to 0°C from −10°C and then see how much energy is left to melt some of the ice.

$$Q = Mc_{ice}\Delta T = (1.00 \, \text{kg})(2090 \, \text{J/kg} \cdot \text{K})(10 \, \text{K}) = 20900 \, \text{J}$$

Now we subtract: $40000 \, \text{J} - 20900 \, \text{J} = 19100 \, \text{J}$; this is the energy left to melt the ice.

$$M = \frac{Q}{L_f} = \frac{19100 \, \text{J}}{3.33 \times 10^5 \, \text{J/kg}} = 0.0574 \, \text{kg}$$

The correct answer is B.

Assess: There was enough energy to raise the water to 0°, but not enough to melt it all.

Problems

P12.3. Prepare: We'll first compute how many moles of hydrogen peroxide molecules there are in 100 g and then use Equation 12.1 to find how many particles that is. The molecular mass number for H_2O_2 is $2\times1+2\times16=34$.

Solve: $(100 \text{ g})(1 \text{ mol/34 g})=2.94$ mol of hydrogen peroxide molecules. However, there are two hydrogen atoms in each molecule of hydrogen peroxide, so there are $2\times2.94 \text{ mol}=5.88$ mol of hydrogen atoms.

$$N=nN_A=(5.88 \text{ mol})(6.02\times10^{23} \text{ mol}^{-1})=3.5\times10^{24}$$

Assess: Three trillion trillion hydrogen atoms is a lot, but one gets used to huge numbers in these types of problems.

P12.5. Prepare: The volume is clearly the product of the three length measurements; the issue is converting the units. First multiply $L\times W\times H$ to get the number of cm^3, then convert to m^3.

Solve:

$$V=(200 \text{ cm})(40 \text{ cm})(3.0 \text{ cm})=24,000 \text{ cm}^3$$

Now remember that while $1 \text{ m}=100 \text{ cm}, 1 \text{ m}^3\neq100 \text{ cm}^3$. Instead, $1 \text{ m}^3=1,000,000 \text{ cm}^3$.

$$24,000 \text{ cm}^3=(24,000 \text{ cm}^3)\left(\frac{1 \text{ m}^3}{1,000,000 \text{ cm}^3}\right)=0.024 \text{ m}^3$$

Assess: The answer is small—not a very big fraction of one cubic meter; however, this is reasonable given the small height. The conversion factor comes from $(1 \text{ m}/100 \text{ cm})^3$.

P12.9. Prepare: Solve $\Delta E_{th}=\frac{3}{2}Nk_B\Delta T$ for ΔT.

Solve:

$$\Delta T=\frac{2}{3}\frac{\Delta E_{th}}{Nk_B}=\frac{2}{3}\frac{-4.3 \text{ J}}{(2.2\times10^{22})(1.38\times10^{-23})}=-9.4 \text{ K}=-9.4 °\text{C}$$

$$T_f=T_i+\Delta T=20°\text{C}-9.4°\text{C}=11°\text{C}$$

Assess: We expected the temperature to drop from the removed thermal energy.

P12.11. Prepare: The absolute pressure is the gauge pressure plus one atmosphere at sea level. 1 atm $=14.7$ psi.

Solve:

$$p=p_g+1 \text{ atm}=35.0 \text{ psi}+14.7 \text{ psi}=49.7 \text{ psi}$$

Assess: The difference between p and p_g is due to the fact that your tire gauge measures pressure *differences*.

P12.17. Prepare: Equation 12.15 applies. We must convert all quantities to SI units.

Solve: Converting units,

$$T=-120+273=153 \text{ K}$$

$$V=(2.0 \text{ L})\left(\frac{10^{-3} \text{ m}^3}{1 \text{ L}}\right)=2.0\times10^{-3} \text{ m}^3$$

Using Equation 12.15,

$$p=\frac{nRT}{V}=\frac{(3.0 \text{ mol})(8.31 \text{ J/(mol}\cdot\text{K)})(153 \text{ K})}{2.0\times10^{-3} \text{ m}^3}=1.9\times10^6 \text{ Pa}$$

The pressure is 1900 kPa.

Assess: Note that this is about twenty times atmospheric pressure.

P12.21. Prepare: The gas is assumed to be ideal and it expands isothermally.
Solve: (a) Isothermal expansion means the temperature stays unchanged. That is $T_f = T_i$.
(b) The before-and-after relationship of an ideal gas under isothermal conditions is

$$\frac{p_i V_i}{T_i} = \frac{p_f V_f}{T_f} \Rightarrow p_f = p_i \frac{V_i}{V_f} = p_i \left(\frac{V_i}{2V_i}\right) = \frac{p_i}{2}$$

Assess: The gas has a lower pressure at the larger volume, as we would expect.

P12.23. Prepare: The isobaric heating means that the pressure of the argon gas stays unchanged. Argon gas in the container is assumed to be an ideal gas. We must first convert the volumes and temperatures to SI units with $V_1 = 50 \text{ cm}^3 = 50 \times 10^{-6} \text{ m}^3$, $T_1 = 20°C = (273 + 20) \text{ K} = 293 \text{ K}$, and $T_2 = 300°C = (300 + 273) \text{ K} = 573 \text{ K}$.
Solve: (a) The container has only argon inside with $n = 0.10$ mol. This produces a pressure

$$p_1 = \frac{nRT_1}{V_1} = \frac{(0.10 \text{ mol})(8.31 \text{ J/(mol·K)})(293 \text{ K})}{50 \times 10^{-6} \text{ m}^3} = 4.87 \times 10^6 \text{ Pa} = 4870 \text{ kPa}$$

An ideal gas process has $p_2 V_2 / T_2 = p_1 V_1 / T_1$. Isobaric heating to a final temperature $T_2 = 300°C = 573 \text{ K}$ has $p_2 = p_1$, so the final volume is

$$V_2 = \frac{p_1}{p_2} \frac{T_2}{T_1} V_1 = 1 \times \frac{573}{293} \times 50 \text{ cm}^3 = 98 \text{ cm}^3$$

(b)

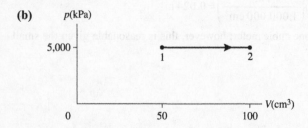

P12.27. Prepare: Assume that the gas is ideal. We will make use of the following conversions: 1 atm $= 1.013 \times 10^5$ Pa and 1 cm$^3 = 1 \times 10^{-6}$ m^3.
Solve: (a) Because the process is at a constant pressure, it is isobaric.
(b) For an ideal gas at constant pressure,

$$\frac{V_2}{T_2} = \frac{V_1}{T_1} \Rightarrow T_2 = T_1 \frac{V_2}{V_1} = [(273 + 900) \text{ K}] \frac{100 \text{ cm}^3}{300 \text{ cm}^3} = 391 \text{ K} \approx 120°C$$

(c) Using the ideal-gas law $p_2 V_2 = nRT_2$,

$$n = \frac{p_2 V_2}{RT_2} = \frac{(3 \times 1.013 \times 10^5 \text{ Pa})(100 \times 10^{-6} \text{ m}^3)}{(8.31 \text{ J/(mol·K)})(391 \text{ K})} = 9.4 \times 10^{-3} \text{ mol}$$

P12.29. Prepare: For a gas in a sealed container (n is constant) we use Equation 12.16.

$$\frac{p_f V_f}{T_f} = \frac{p_i V_i}{T_i}$$

In this case we want to solve for T_f,

$$T_f = \frac{p_f V_f}{p_i V_i} T_i$$

but we get to cancel the volumes since they are the same.
We need to take the usual steps of converting temperatures to the absolute scale. The problem states that the pressures given are already absolute pressures.

$$T_i = 0.00°C = 273 \text{ K}$$

Solve:

$$T_f = \frac{p_f}{p_i}T_i = \frac{65.1\,\text{kPa}}{55.9\,\text{kPa}}273\,\text{K} = 318\,\text{K} = 45°\text{C}$$

Assess: We expected the final temperature to be higher than the initial temperature as the pressure rose. The answer is a reasonable real-life temperature.

P12.33. Prepare: We are given much of the data needed in Equation 12.21 except the coefficient of linear expansion for steel, which we look up in Table 12.3. $\Delta L = 0.73$ mm, $\Delta T = 13$ K, $\alpha_{\text{steel}} = 12 \times 10^{-6}\,\text{K}^{-1}$.

We want to know the original length, so we solve Equation 12.21 for that quantity.

Solve:

$$L_i = \frac{\Delta L}{\alpha \Delta T} = \frac{(0.73\,\text{mm})}{(12 \times 10^{-6}\,\text{K}^{-1})(13\,\text{K})}\left(\frac{1\,\text{m}}{1000\,\text{mm}}\right) = 4.7\,\text{m}$$

Assess: This seems to be a reasonable answer—in the realm of daily life, about the width of a room. It would have taken an aluminum beam only about half that long to produce the same ΔL under the same ΔT.

P12.35. Prepare: The coefficient of thermal expansion works for any linear dimension, such as the diameter of a two-dimensional hole. Equation 12.19 and Table 12.3 apply. The linear coefficient for steel is $12 \times 10^{-6}\,\text{K}^{-1}$.

Solve: The expansion of the hole

$$\Delta L = \alpha L_i \Delta T \Rightarrow \Delta T = \frac{\Delta L}{\alpha L_i} = \frac{0.003\,\text{cm}}{(12 \times 10^{-6}\,\text{K}^{-1})(2.000\,\text{cm})} = 130\,\text{K}$$

Assess: If it seems counterintuitve that the hole should grow, pretend it was filled with steel plate.

P12.37. Prepare: We need to rearrange Equation 12.2 to give the fractional volume expansion (the percentage expansion is then just 100 times that fractional expansion).

Look up $\beta_{\text{Al}} = 69 \times 10^{-6}\,\text{K}^{-1}$ in Table 12.3. $\Delta T = 120$ K.

Solve:

$$\frac{\Delta V}{V_i} = \beta \Delta T = (69 \times 10^{-6}\,\text{K}^{-1})(120\,\text{K}) = 0.0083 = 0.83\%$$

Assess: The expansion is small, but greater than it would have been for steel. While 120°C is a larger temperature swing than we might see on a daily basis, there are situations (engines, etc.) where there are significant temperature differences, and people who design and build precise things must take such expansions into account.

P12.39. Prepare: The mass of the mercury is $M = 20$ g $= 2.0 \times 10^{-2}$ kg, the specific heat from Table 12.4 $c_{\text{mercury}} = 140$ J/kg K, the boiling point $T_b = 357°\text{C}$, and the heat of vaporization from Table 12.5 $L_V = 2.96 \times 10^5$ J/kg. We will use Equation 12.26 for obtaining heat needed to raise its temperature to boiling and Equation 12.26 for obtaining heat needed to boil mercury into vapors at the boiling temperature. Note that heating the mercury at its boiling point changes its thermal energy without a change in temperature. We also note that ΔT is the same whether we calculate it in Kelvins or in °C, so we don't have to convert °C into K.

Solve: The heat required for the mercury to change to the vapor phase is the sum of two steps. The first step is

$$Q_1 = Mc_{\text{mercury}}\Delta T = (2.0 \times 10^{-2}\,\text{kg})(140\,\text{J/(kg·K)})(357°\text{C} - 20°\text{C}) = 944\,\text{J}$$

The second step is

$$Q_2 = ML_V = (2.0 \times 10^{-2}\,\text{kg})(2.96 \times 10^5\,\text{J/kg}) = 5920\,\text{J}$$

The total heat needed is 6870 J, which will be reported as 6900 J.

Assess: More energy is needed to vaporize mercury (86%) than to warm it to its boiling temperature (14%), as we would expect.

P12.43. Prepare: We are asked to find the time it takes for the alligator's temperature to rise from 25°C to 30°C, that is by 5 K. It is absorbed at a rate of 1200 W. We use the mammalian specific heat: $c = 3400$ J/(kg·K). We will need to use the definition of power, which in this case can be written as $P = Q/\Delta t$, as well as the formula for heat absorbed in connection with an increase in temperature, $Q = Mc\Delta T$.

Solve: The time needed for the alligator to reach its final temperature can be obtained by solving the power equation: $\Delta t = Q/P$. This, coupled with the formula for heat gives us the following:

$$\Delta t = \frac{Q}{P} = \frac{Mc\Delta T}{P} = \frac{(300 \text{ kg})(3400 \text{ J/(kg} \cdot \text{K)})(5 \text{ K})}{1200 \text{ W}} = 4250 \text{ s} = 71 \text{ min}$$

This is just over 1 hour.

Assess: This seems like a reasonable length of time. The alligator will still have most of the day at the warmer body temperature to catch prey.

P12.45. Prepare: We need to find the temperature increase and we are given the metabolic power $P = 1000$ W, the time of exercise $\Delta t = 30$ min $= (30 \text{ min})(60 \text{ s})/(1 \text{ min}) = 1800$ s, and the man's mass $m = 70$ kg. We need the formula for power, which in this case is $P = \Delta E_{th}/\Delta t$ and Equation 12.22 for heat absorbed. Here the heat produced by the exercise goes to increasing the thermal energy of the man, so we can write the following: $\Delta E_{th} = Mc\Delta T$. For the specific heat, we use the mammalian specific heat from Table 12.4: $c = 3400$ J/kg $\cdot$ K.

Solve: First we will find the thermal energy produced by the exercise by solving the power equation.

$$\Delta E_{th} = P\Delta t = (1000 \text{ W})(1800 \text{ s}) = 1.8 \times 10^6 \text{ J}$$

We can solve the equation $\Delta E_{th} = Mc\Delta T$ for ΔT.

$$\Delta T = \Delta E_{th}/Mc = \frac{(1.8 \times 10^6 \text{ J})}{(70 \text{ kg})(3400 \text{ J/(kg} \cdot \text{K)})} = 7.56 \text{ K}$$

To two significant figures, his temperature increases by 7.6 K which is 14°F.

Assess: Even though he only exercised for 30 min, the man's temperature has increased by 14°F. This would bring a temperature of 98.6°F up to about 113°F, which would be very dangerous.

P12.49. Prepare: We have a thermal interaction between the aluminum pan and the water that leads to the common final temperature denoted by T_f. The initial temperature of the aluminum pan is unknown, but that of the water is known. The specific heats of aluminum and water from Table 12.4 are $c_{Al} = 900$ J/(kg $\cdot$ K) and $c_w = 4190$ J/(kg $\cdot$ K). The mass of the aluminum pan is $M_{Al} = 0.750$ kg and the mass of water is $M_{water} = 10$ kg. We can determine the initial temperature of the pan using Equations 12.24 and 12.27.

Solve: The conservation of energy equation $Q_{Al} + Q_{water} = 0$ J is

$$M_{Al}c_{Al}(T_f - T_{i \text{ Al}}) + M_{water} \, c_{water}(T_f - T_{i \text{ water}})$$

The pan and water reach a common final temperature $T_f = 24.0°C$

$$(0.750 \text{ kg})(900 \text{ J/(kg} \cdot \text{K)})(24.0°C - T_{i \text{ Al}}) + (10 \text{ kg/m}^3)(4190 \text{ J/(kg} \cdot \text{K)})(24.0°C - 20.0°C)$$
$$= (675.0 \text{ J/K})(24.0°C - T_{i \text{ Al}}) + 167,600 \text{ J} = 0 \text{ J} \Rightarrow T_{i \text{ Al}} = 272°C$$

Assess: Due to (1) the large specific heat of water compared to aluminum and (2) the large mass of water compared to aluminum, an increase of temperature of 4°C needed high initial temperature of the pan.

P12.51. Prepare: This is a basic calorimetry problem. We are given the mass and initial temperature of the aluminum $m_{Al} = 200$ g and $T_{i \text{ Al}} = -20°C$, as well as the mass and initial temperature of the coffee $m_c = 500$ g and $T_{i \text{ c}} = 85°C$. We obtain the specific heats from Table 12.4, according to which $c_{Al} = 900$ J/(kg $\cdot$ K) and $c_c = 4190$ J/(kg $\cdot$ K).

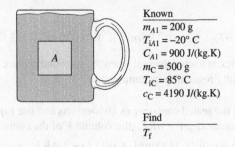

Known
$m_{A1} = 200$ g
$T_{iA1} = -20°$ C
$C_{A1} = 900$ J/(kg·K)
$m_C = 500$ g
$T_{iC} = 85°$ C
$c_C = 4190$ J/(kg·K)

Find
T_f

Solve: The basic calorimetry equation becomes $m_{A1}c_{A1}\Delta T_{A1} + m_c c_c \Delta T_c = 0$ which we can solve for the final temperature:

$$(0.200 \text{ kg})(900 \text{ J/kg})(T_f - -20°C) + (0.500 \text{ kg})(4190 \text{ J/(kg·K)})(T_f - 85°C) = 0 \text{ J}$$
$$\Rightarrow (2280 \text{ J/K})T_f - 174000 \text{ J} = 0$$
$$\Rightarrow T_f = 76°C$$

Assess: Notice that even though the aluminum was comparable in mass to the coffee, its temperature change of $96°C$ was far more than the temperature change of the coffee, $9°C$. This is a result of water having a much higher specific heat than metal.

P12.57. Prepare: From the first law of thermodynamics, $Q = \Delta E_{th} - W$, where W is the work done by the gas. $W = 0$ at constant volume, so, using Equation 12.28, $\Delta E_{th} = Q = nC_V\Delta T$. From Table 12.6, the value of C_V for a monatomic gas is 12.5 J/(mol·K) (which is equal to $3R/2$). For a diatomic gas, we take C_V to be 20.8 J/(mol·K) (which is equal to $5R/2$).
Solve: For a monatomic gas,

$$\Delta E_{th} = nC_V\Delta T = 1.0 \text{ J} = (1.0 \text{ mol})(12.5 \text{ J/(mol·K)})\Delta T \Rightarrow \Delta T = 0.0800°C \text{ or } 0.080 \text{ K}$$

P12.63. Prepare: The rate of energy loss by radiation is given by Equation 12.35.

$$\frac{Q}{\Delta t} = e\sigma A T^4$$

We are given $e = 0.20$, $T = 700°C = 973$ K, $A = 6 \times (2.0 \text{ cm} \times 2.0 \text{ cm}) = 24 \text{ cm}^2 = 0.0024 \text{ m}^2$. The textbook gives Stefan's constant as $\sigma = 5.67 \times 10^{-8}$ W/(m^2·K^4).
Solve:

$$\frac{Q}{\Delta t} = e\sigma A T^4 = (0.20)(5.67 \times 10^{-8} \text{ W/(m}^2 \cdot \text{K}^4))(0.0024 \text{ m}^2)(973 \text{ K})^4 = 24 \text{ W}$$

Assess: $700°C$ is quite hot, so the cube radiates a reasonable amount of energy, but even at $700°C$ the radiation is mostly infrared, not visible. If we double the (absolute) temperature the total radiation would increase by a factor of 16 (due to the T^4) and also a greater portion of the radiation would be in the visible range.

P12.65. Prepare: The rate of net energy loss by radiation is given by Equation 12.36.

$$\frac{Q_{net}}{\Delta t} = e\sigma A(T^4 - T_0^4)$$

where T_0 is the termperature of the surroundings.

We are given $T = 30°C = 303$ K, $T_0 = -10°C = 263$ K, and $A = 0.030 \text{ m}^2$. We are told to assume the emissivity of seal skin is the same as human skin; the text gives this vaule as $e = 0.97$.
The textbook gives Stefan's constant as $\sigma = 5.67 \times 10^{-8}$ W/(m^2·K^4).

Solve:

$$\frac{Q_{net}}{\Delta t} = e\sigma A(T^4 - T_0^4) = (0.97)(5.67 \times 10^{-8} \text{ W/(m}^2 \cdot \text{K}^4))(0.030 \text{ m}^2)[(303 \text{ K})^4 - (263 \text{ K})^4] = 6.0 \text{ W}$$

Assess: 6 W isn't a lot, but it is sufficient to cool the seal when the surroundings are very cool. If there were no thermal windows the seal would have difficulty regulating its temperature.

P12.71. Prepare: Treat the gas in the sealed container as an ideal gas and use Equation 12.16.
Solve: (a) From the ideal gas law equation $pV = nRT$, the volume V of the container is

$$V = \frac{nRT}{p} = \frac{(2.0 \text{ mol})(8.31 \text{ J/(mol} \cdot \text{K}))[(273 + 30) \text{ K}]}{1.013 \times 10^5 \text{ Pa}} = 0.050 \text{ m}^3 = 50 \text{ L}$$

(b) The before-and-after relationship of an ideal gas in a sealed container (constant volume) is

$$\frac{p_1 V}{T_1} = \frac{p_2 V}{T_2} \Rightarrow p_2 = p_1 \frac{T_2}{T_1} = (1.0 \text{ atm})\frac{(273 + 130) \text{ K}}{(273 + 30) \text{ K}} = 1.3 \text{ atm}$$

Assess: Note that gas-law calculations *must* use T in kelvins and pressure *must* be in Pa.

P12.73. Prepare: Treat the helium gas in the sealed cylinder as an ideal gas. The volume of the cylinder is

$$V = \pi r^2 h = \pi (0.05 \text{ m})^2 (0.30 \text{ m}) = 2.356 \times 10^{-3} \text{ m}^3$$

The gauge pressure of the gas is

$$120 \text{ psi} \times \frac{1 \text{ atm}}{14.7 \text{ psi}} \times \frac{1.013 \times 10^5 \text{ Pa}}{1 \text{ atm}} = 8.269 \times 10^5 \text{ Pa}$$

so the absolute pressure of the gas is $8.269 \times 10^5 \text{ Pa} + 1.013 \times 10^5 \text{ Pa} = 9.282 \times 10^5 \text{ Pa}$. The temperature of the gas is $T = (273 + 20) \text{ K} = 293 \text{ K}$.
Solve: (a) The number of moles of the gas in the cylinder is

$$n = \frac{pV}{RT} = \frac{(9.282 \times 10^5 \text{ Pa})(2.356 \times 10^{-3} \text{ m}^3)}{(8.31 \text{ J/(mol} \cdot \text{K}))(293 \text{ K})} = 0.898 \text{ mol}$$

So, the number of atoms is

$$N = nN_A = (0.898 \text{ mol})(6.02 \times 10^{23} \text{ mol}^{-1}) = 5.4 \times 10^{23} \text{ atoms}$$

(b) The mass of the helium is

$$M = nM_{mol} = (0.898 \text{ mol})(4 \text{ g/mol}) = 3.6 \text{ g}$$

P12.79. Prepare: Assume that the compressed air in the cylinder is an ideal gas. The volume of the air in the cylinder is a constant. We will use Equation 12.16 to calculate the new pressure in atm and compare it with the maximum pressure (in atm) of the compressed gas that the cylinder can withstand.
Solve: Using the before-and-after relationship of an ideal gas,

$$\frac{p_f V_f}{T_f} = \frac{p_i V_i}{T_i} \Rightarrow p_f = p_i \frac{T_f}{T_i} \frac{V_i}{V_f} = (25 \text{ atm})\left(\frac{1223 \text{K}}{293 \text{K}}\right)\frac{V_i}{V_i} = 104 \text{ atm}$$

where we have converted to the Kelvin temperature scale. Because the pressure does not exceed 110 atm, the compressed air cylinder does not blow.

P12.83. Prepare: Assume the gas is an ideal gas. We will use Equation 12.15, convert all quantities to SI units, and refer to Figure P12.83 for initial and final pressure and volume.
Solve: (a) We can find the temperatures directly from the ideal gas law.

$$T_1 = \frac{p_1 V_1}{nR} = \frac{(3.0 \text{ atm} \times 101,300 \text{ Pa/atm})(1000 \text{ cm}^3 \times 10^{-6} \text{ m}^3/\text{cm}^3)}{(0.10 \text{ mol})(8.31 \text{ J/(mol} \cdot \text{K}))} = 366 \text{ K} = 93^{\circ}\text{C}$$

$$T_2 = \frac{p_2 V_2}{nR} = \frac{(1.0 \text{ atm} \times 101,300 \text{ Pa/atm})(3000 \text{ cm}^3 \times 10^{-6} \text{ m}^3/\text{cm}^3)}{(0.10 \text{ mol})(8.31 \text{ J/(mol} \cdot \text{K}))} = 366 \text{ K} = 93^{\circ}\text{C}$$

(b) $T_2 = T_1$, so this is an isothermal process.

(c) A constant volume process has $V_3 = V_2$. Because $p_1 = 3p_2$, restoring the pressure to its original value means that $p_3 = 3p_2$. From the ideal gas law,

$$\frac{p_3 V_3}{T_3} = \frac{p_2 V_2}{T_2} \Rightarrow T_3 = \left(\frac{p_3}{p_2}\right)\left(\frac{V_3}{V_2}\right)T_2 = 3 \times 1 \times T_2 = 3 \times 366\ \text{K} = 1098\ \text{K} = 825°\text{C}$$

The temperature to be reported is thus 830°C.

P12.85. Prepare: Please refer to the following figure. The work done in expanding the swim bladder is given by $W = p\Delta V$. To use this, we need to find the pressure at the lower depth. The temperature and number of moles of gas remain constant as the fish descends, so from the ideal gas law, pV is constant, that is $p_i V_i = p_f V_f$ or, rearranging factors, $p_f = \dfrac{p_i V_i}{V_f}$. Since the ratio of the final volume to the initial volume is 60% and the initial pressure was 3.0 atm, we can say that $V_f = (0.60)V_i$ and $p_f = \dfrac{p_i V_i}{V_f} = \dfrac{(3.0\ \text{atm})V_i}{(0.60)V_i} = 5.0\ \text{atm}$. We also are given: $V_i = 5.0 \times 10^{-4}\ \text{m}^3$, from which it follows that $V_f = (0.60)V_i = 3.0 \times 10^{-4}\ \text{m}^3$.

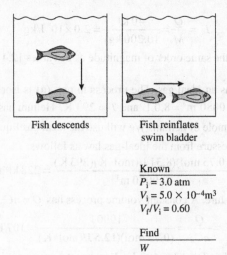

Fish descends   Fish reinflates
swim bladder

Known
$P_i = 3.0$ atm
$V_i = 5.0 \times 10^{-4}\text{m}^3$
$V_f/V_i = 0.60$

Find
W

Solve: After descending, the fish pumps gas into its swim bladder to go from the present volume of $3.0 \times 10^{-4}\ \text{m}^3$ back to its original volume of $5.0 \times 10^{-4}\ \text{m}^3$. The work done is as follows:

$$W = p\Delta V = (5.0\ \text{atm})\left(\frac{1.013 \times 10^5\ \text{Pa}}{1\ \text{atm}}\right)(5.0 \times 10^{-4}\ \text{m}^3 - 3.0 \times 10^{-4}\ \text{m}^3) = 100\ \text{J}$$

Assess: The units work out to joules since $1\ \text{Pa} = 1\ \text{N/m}^2$.

P12.87. Prepare: Since the compression is adiabatic, there is no heat flow and the first law of thermodynamics says that $\Delta E_{\text{th}} = W$ where W is the work done on the gas. So we need to find the change in thermal energy from Equation 12.32.
Solve: Combining the two equations mentioned above, we have the following:

$$W = \Delta E_{\text{th}} = \tfrac{3}{2}nR\Delta T = \tfrac{3}{2}(0.030\ \text{mol})(8.315\ \text{J/(mol}\cdot\text{K)})(50°\text{C} - 10°\text{C}) = 15\ \text{J}$$

Assess: We were able to use degrees Celsius in the above calculation since we were subtracting two temperatures.

P12.91. Prepare: From Table 11.4, the metabolic power of a 68 kg cyclist is 480 W. We assume that 25% of this goes to propelling the cyclist and the other 75%, or 360 W, becomes heat which serves to evaporate perspiration. Equation 12.28 gives the heat needed to evaporate a liquid. From the discussion following Table 12.5, a good value for the latent heat of vaporization of sweat is $L_v = 2.4 \times 10^6\ \text{J/kg}$.

Solve: We solve Equation 12.28 for m and get $m = Q/L_v$. This is the mass of perspiration which heat Q could evaporate. We know $\dfrac{Q}{\Delta T} = 360$ W and we can combine this with the preceding equation to obtain the following:

$$\frac{m}{\Delta t} = \frac{Q/L_v}{\Delta t} = \frac{Q/\Delta t}{L_v} = \frac{360 \text{ J/s}}{2.4 \times 10^6 \text{ J/kg}} = 1.5 \times 10^{-4} \text{ kg/s} = 0.54 \text{ kg/h}$$

Assess: A kilogram of water has a volume of about one liter, so this is about half a liter per hour. A value on the order of one liter per hour seems reasonable.

P12.95. Prepare: Heating the material increases its thermal energy. The material melts at 300°C and undergoes a solid-liquid phase change. The material's temperature increases from 300°C to 1500°C. Boiling occurs at 1500°C and the material undergoes a liquid-gas phase change. We will use Equations 12.23 and 12.24 to determine the specific heat and the heat of vaporization of the liquid.

Solve: (a) In the liquid phase, the specific heat of the liquid can be obtained as follows:

$$\Delta Q = Mc\Delta T \Rightarrow c = \frac{1}{M}\frac{\Delta Q}{\Delta T} = \left(\frac{1}{0.200 \text{ kg}}\right)\left(\frac{20 \text{ kJ}}{1200 \text{ K}}\right) = 83 \text{ J/(kg} \cdot \text{K)}$$

(b) The latent heat of vaporization is

$$L_v = \frac{Q}{M} = \frac{40 \text{ kJ}}{(0.200 \text{ kg})} = 2.0 \times 10^5 \text{ J/kg}$$

Assess: The values obtained are of the same order of magnitude as in Tables 12.4 and 12.5 for a few materials.

P12.101. Prepare: Treat helium as an ideal gas. The process in part **(a)** is isochoric and the process in part **(b)** is isobaric. Initially $V_i = (0.20 \text{ m})^3 = 0.0080 \text{ m}^3 = 8.0$ L and $T_i = 293$ K. Helium has an atomic mass number $A = 4$, so 3 g of helium is $n = M/M_{mol} = 0.75$ mole of helium. We will use the ideal gas Equations 12.15 and 12.14.

Solve: (a) We can find the initial pressure from the ideal-gas law as follows:

$$p_i = \frac{nRT_i}{V_i} = \frac{(0.75 \text{ mol})(8.31 \text{ J/(mol} \cdot \text{K)})(293 \text{ K})}{0.0080 \text{ m}^3} = 228 \text{ kPa} = 2.25 \text{ atm}$$

Heating the gas will raise its temperature. A constant volume process has $Q = nC_V\Delta T$, so

$$\Delta T = \frac{Q}{nC_V} = \frac{1000 \text{J}}{(0.75 \text{ mol})(12.5 \text{ J/(mol} \cdot \text{K)})} = 107 \text{ K}$$

This raises the final temperature to $T_f = T_i + \Delta T = 400$ K. Because the process is isochoric,

$$\frac{p_f}{T_f} = \frac{p_i}{T_i} \Rightarrow p_f = \frac{T_f}{T_i}p_i = \frac{400 \text{ K}}{293 \text{ K}}(2.25 \text{ atm}) = 3.1 \text{ atm}$$

(b) The initial conditions are the same as in part **(a)**, but now $Q = nC_p\Delta T$. Thus,

$$\Delta T = \frac{Q}{nC_p} = \frac{1000 \text{ J}}{(0.75 \text{ mol})(20.8 \text{ J/(mol} \cdot \text{K)})} = 64.1 \text{ K}$$

Now the final temperature is $T_f = T_i + \Delta T = 357$ K. Because the process is isobaric,

$$\frac{V_f}{T_f} = \frac{V_i}{T_i} \Rightarrow V_f = \frac{T_f}{T_i}V_i = \frac{357 \text{ K}}{293 \text{ K}}(0.0080 \text{ m}^3) = 0.00975 \text{ m}^3 = 9.75 \text{ L}$$

which is to be reported as 9.8 L.

P12.105. Prepare: The equation relating the rate of thermal energy transfer due to conduction (the thing we want) and the thermal conductivity is given by Equation 12.34. $\Delta T = 30$ K, $A = 370$ m^2.

Solve:

$$\frac{Q}{\Delta t} = \left(\frac{kA}{L}\right)\Delta T = \frac{(0.040 \text{ W/m} \cdot \text{K})(370 \text{ m}^2)}{0.10 \text{ m}}(30 \text{ K}) = 4.4 \text{ kW}$$

13

FLUIDS

Q13.3. Reason: If the chunk is heavy (dense) enough to sink in water you would put the chunk into a known volume of water in a graduated cylinder and note the rise; this would give the volume of the chunk. A simple measurement of the chunk's mass on a pan balance would then allow you to use $\rho = m/V$.

Assess: If the chunk floats in water then you would have to find a way to submerse it to find the volume.

Q13.5. Reason: It all has to do with pressure increasing with depth, which is a result of the gravitational force. If, when receiving a transfusion, the bag were held below your body the blood would not flow up into your body, and if, when donating blood, the bag were held above your body, no blood would flow up into it. The blood wants to flow down.

Assess: It is kind of neat to notice things like this and think physics thoughts, even while getting a transfusion or donating blood!

Q13.7. Reason: The pressure only depends on the depth from the opening. Since A, B, and C are all at the same depth below the opening at E the pressure is the same for each.

$$p_A = p_B = p_C$$

Assess: While there is a taller column of water over B, you can think that the top of the container at D and F pushes down on the fluid too, so the pressure at A, B, and C is the same.

Q13.11. Reason: The area of the bottom of the column is 1m^2. The pressure there is $1 \text{ atm} = 101.3 \text{ kPa} = 101.3 \text{ kN/m}^2$. So the air is pushing down with a force of 101.3 kN which is its weight.

Assess: This is a large force. That's why it can collapse a steel barrel if the air is sucked out of the inside.

Q13.13. Reason: Two points on a horizontal line, connected by a single liquid in hydrostatic equilibrium, are at the same pressure. So points A and B are at the same pressure.

Assess: You might imagine that point A has to support the weight of the ship, while point B does not. But remember that the ship has displaced a weight of water equal to the ship's weight, so the total weight supported by point A is unchanged by the presence of the ship.

Q13.15. Reason: The fraction of the object below the surface of the liquid is the object's density as a fraction of the liquid's density (if we can ignore the density of the air). Since A has the greatest fraction below the surface it is the most dense. The least dense (B) floats with the largest fraction of its volume above the fluid level (the smallest fraction below the liquid level).

$$\rho_A > \rho_C > \rho_B$$

Assess: You've heard that only 10% of an iceberg is visible above the surface of the ocean. That means 90% of the iceberg is below the surface. Therefore the density of ice is 90% the density of seawater. Verify this by looking in Table 13.1 and finding on the Web the density of ice, $\rho_{ice} = 917 \text{ kg/m}^3$.

Q13.19. Reason: The air in the room produces an upward buoyant force on you. As the pressure in the room decreases so does the density of air, so the buoyant force decreases. The scale reading will increase.
Assess: Note that the scale is surrounded by air so the decrease in pressure does not change the force of the air on the scale.

Q13.21. Reason: Adding salt to the water increases its density. When the density of the water matches that of the egg, the egg becomes neutrally buoyant and floats.
Assess: This makes sense. An egg consists essentially of water mixed with the substance of the embryo, which makes the density of the egg slightly larger than that of water.

Q13.25. Reason: The force that the block exerts on the water is equal to the buoyant force of the water on the block, which is exactly equal to the weight of the water displaced. The weight of the water that has run off is matched by the force the block exerts on the water. The scale reading remains the same.
Assess: Note that the block must be floating for the reading to remain the same.

Q13.27. Reason: The lower the velocity of fluid, the higher the pressure at a given point in the pipe. The velocities from highest to smallest are $v_2 > v_1 > v_3$.
Assess: This can be seen from Bernoulli's principle, Equation 13.14. The kinetic energy of a fluid element balances with the work done by the pressure force.

Q13.31. Reason: The pressure is reduced at the chimney due to the movement of the wind above. Thus, the air will flow in the window and out the chimney.
Assess: Prairie dogs ventilate their burrows this way; a small breeze above their mound lowers the pressure there and allows the air in the burrow to move between openings of different types or heights.

Q13.35. Reason: Since the question talks about the "extra pressure" we will ignore the air pressure above the water; you have an equal amount of air pressure on the inside. The 7 N quoted is the "increased pressure." Therefore the equation we need is Equation 13.5 without the p_0 term,

$$p = \rho g d$$

where we want to solve for d. We will also use $p = F/A$ and round g to 10 m/s^2 for one significant figure accuracy.

$$d = \frac{p}{\rho g} = \frac{F/A}{\rho g} = \frac{7 \text{ N}/(7 \times 10^{-5} \text{ m}^2)}{(1000 \text{ kg/m}^3)(10 \text{ m/s}^2)} = 10 \text{ m}$$

The correct answer is D.
Assess: This one significant figure calculation can easily be done in the head without a calculator. The answer seems plausible, and the other choices seem too small. The units cancel appropriately to leave the answer in m.

Q13.39. Reason: We'll use the equation of continuity, Equation 13.12, and make all the assumptions inherent in that equation (no leakage, etc.). We'll also use $A = \pi r^2$. Call the plunger point 1 and the nozzle point 2, and solve for v_1.

$$v_1 = \frac{v_2 A_2}{A_1} = \frac{v_2 \pi r_2^2}{\pi r_1^2} = \frac{v_2 r_2^2}{r_1^2} = \frac{(10 \text{ m/s})(1.0 \times 10^{-3} \text{ m})^2}{(1.0 \times 10^{-2} \text{ m})^2} = 0.10 \text{ m/s}$$

The correct answer is B.
Assess: The answer seems to be a reasonable slow and steady pace. The ratio of the speeds is the square of the ratio of the radii. The last step (the computation) can easily be done without a calculator.

Problems

P13.1. Prepare: In SI Units $1 \text{ L} = 10^{-3} \text{ m}^3$ and $1 \text{ g} = 10^{-3} \text{ kg}$.
Solve: The density of the liquid is

$$\rho = \frac{m}{V} = \frac{0.120 \text{ kg}}{100 \text{ mL}} = \frac{0.120 \text{ kg}}{100 \times 10^{-3} \times 10^{-3} \text{ m}^3} = 1200 \text{ kg/m}^3$$

Assess: The liquid's density is a little more than that of water (1000 kg/m^3) and is a reasonable number.

P13.3. Prepare: The volume of the sphere will be reduced by a factor of 8 when its radius is halved. We will use the definition of mass density $\rho = m/V$.

Solve: The new density is

$$\rho' = \frac{m}{V/8} = 8\frac{m}{V} = 8\rho = 8(1.4 \text{ kg/m}^3) = 11 \text{ kg/m}^3$$

Assess: If the mass is constant and the volume is reduced by a factor of 8, the density will increases by a factor of 8.

P13.9. Prepare: Assume both liquids are incompressible. The densities are given in Table 13.1: 1000 kg/m^3 for water and 900 kg/m^3 for oil. The pressure at the bottom of a liquid is due to the air pressure at the top plus the pressure due to the liquid. In this case the pressure at the top of the water is the pressure at the bottom of the oil, so we apply the concept twice. One further point: since we are asked for gauge pressure we will subtract the atmospheric pressure from the absolute pressure.

Solve: The pressure below the surface can be found from Equation 13.5 as follows:

$$p_{\text{guage}} = p - 1 \text{ atm} = (p_0 + \rho_{\text{w}}gd_{\text{w}} + \rho_{\text{oil}}gd_{\text{oil}}) - 1 \text{ atm} = \rho_{\text{w}}gd_{\text{w}} + \rho_{\text{oil}}gd_{\text{oil}}$$

$$= (1000 \text{ kg/m}^3)(9.80 \text{ m/s}^2)(0.25 \text{ m}) + (900 \text{ kg/m}^3)(9.80 \text{ m/s}^2)(0.15 \text{ m})$$

$$= 3.8 \text{ kPa}$$

Assess: If the water and oil had somehow managed to mix together the answer would still be the same.

P13.11. Prepare: We'll use Equation 13.5, which gives the pressure of a fluid as a function of depth. We know that the pressure at a given depth depends on the depth but not the diameter of the container; however, we use $d = 5.0 \text{ cm}$ to compute the area of the bottom because $F = pA$. $A = \pi r^2 = \pi(2.5 \text{ cm})^2 = 19.63 \text{ cm}^2$.

Solve:

$$F = (p_0 + \rho gd)A$$

$$pA = [101.3 \text{ kPa} + (1000 \text{ kg/m}^3)(9.8 \text{ m/s}^2)(0.35 \text{ m})](19.63 \text{ cm}^2) = [101\,300 \text{ Pa} + 3430 \text{ Pa}](19.63 \text{ cm}^2)$$

$$= [104\,730 \text{ Pa}](19.63 \times 10^{-4} \text{ m}^2) = 210 \text{ N}$$

Assess: The weight of the water added a little bit to the weight of the air; you'd have to go down to a depth of almost 11 m to have the contribution due to the weight of the water equal the contribution due to the weight of the air.

P13.13. Prepare: The density of seawater $\rho_{\text{seawater}} = 1030 \text{ kg/m}^2$.

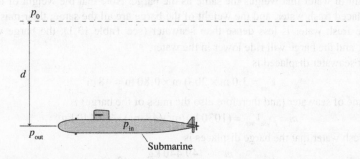
Submarine

Solve: The pressure outside the submarine's window is $p_{\text{out}} = p_0 + \rho_{\text{seawater}}gd$, where d is the maximum safe depth for the window to withstand a force F. This force is $F/A = p_{\text{out}} - p_{\text{in}}$, where A is the area of the window. With $p_{\text{in}} = p_0$, we simplify the pressure equation to

$$p_{out} - p_0 = \frac{F}{A} = \rho_{seawater}gd \Rightarrow d = \frac{F}{A\rho_{seawater}g} \qquad d = \frac{1.0 \times 10^6 \, \text{N}}{\pi(0.10 \, \text{m})^2(1030 \, \text{kg/m}^2)(9.8 \, \text{m/s}^2)} = 3153 \, \text{m} = 3.2 \, \text{km}$$

Assess: A force of 1.0×10^6 N corresponds to a pressure of

$$p = \frac{F}{A} = \frac{1.0 \times 10^6 \, \text{N}}{\pi(0.10 \, \text{m})^2} = 310 \, \text{atm}$$

A depth of 3.2 km is therefore reasonable.

P13.15. Prepare: We assume that there is a perfect vacuum inside the cylinders with $p = 0$ Pa. We also assume that the atmospheric pressure in the room is 1 atm. The flat end of each cylinder has an area $A = \pi r^2 = \pi(0.30 \, \text{m})^2 = 0.283 \, \text{m}^2$.

Solve: (a) The force on each end is

$$F_{atm} = p_0 A = (1.013 \times 10^5 \, \text{Pa})(0.283 \, \text{m}^2) = 2.86 \times 10^4 \, \text{N} = 2.9 \times 10^4 \, \text{N}$$

(b) The net vertical force on the lower cylinder when it is on the verge of being pulled apart is

$$\Sigma F_y = F_{atm} - w_{players} = 0 \, \text{N} \Rightarrow w_{players} = F_{atm} = 2.86 \times 10^4 \, \text{N} \Rightarrow \text{number of players} = \frac{2.86 \times 10^4 \, \text{N}}{(100 \, \text{kg})(9.8 \, \text{m/s}^2)} = 29.2$$

That is, 30 players are needed to pull the two cylinders apart.

Assess: This problem does an excellent job of helping you understand the ramifications of the fact that we live at the bottom of a very deep air ocean (i.e. atmospheric pressure).

P13.21. Prepare: Review Example 13.1. Since the system is closed to the air, we will ignore the atmospheric pressure and simply use $\Delta p = \rho g \Delta d$. We will assume that when you are lying down your brain is at the same level as your heart, so that it is 40 cm above it when you stand.

Solve:

$$\Delta p = \rho g \Delta d = (1060 \, \text{kg/m}^3)(9.8 \, \text{m/s}^2)(0.40 \, \text{m}) = 4155 \, \text{Pa}\left(\frac{1 \, \text{atm}}{101300 \, \text{Pa}}\right)\left(\frac{760 \, \text{mm Hg}}{1 \, \text{atm}}\right) = 31.2 \, \text{mm Hg}$$

Subtract the difference from the original to get the final value, $120 \, \text{mm Hg} - 31 \, \text{mm Hg} = 89 \, \text{mm Hg}$. This is in the range considered low.

Assess: Most people have experienced this a few times when they have stood quickly. If you stand slowly it gives the blood vessels time to constrict and keep the pressure up.

P13.23. Prepare: We know that the barge displaces an amount of seawater that weighs the same as the barge. This will allow us to compute the weight of the barge. Then, as the barge moves into the fresh water we know that it will also displace an amount of water that weighs the same as the barge. Note that the weight of the displaced seawater, the weight of the displaced fresh water, and the weight of the barge are all the same; their masses are also equal, call that value m. Because fresh water is less dense than seawater (see Table 13.1), the barge will displace a greater volume of fresh water, and the barge will ride lower in the water.

Solve: The volume of seawater displaced is

$$V_{sea} = 3.0 \, \text{m} \times 20.0 \, \text{m} \times 0.80 \, \text{m} = 48 \, \text{m}^3$$

The mass of that volume of seawater (and therefore also the mass of the barge) is

$$m = \rho_{sea}V_{sea} = (1030 \, \text{kg/m}^3)(48 \, \text{m}^3) = 49\,440 \, \text{kg}$$

Now, the volume of fresh water that the barge displaces is

$$V_{fresh} = \frac{m}{\rho_{fresh}} = \frac{49\,440 \, \text{kg}}{1000 \, \text{kg/m}^3} = 49.44 \, \text{m}^3$$

Lastly, since the area of the barge has not changed, we solve for the new depth.

$$d = \frac{V_{fresh}}{A} = \frac{49.44 \, \text{m}^3}{3.0 \, \text{m} \times 20.0 \, \text{m}} = 0.824 \, \text{m} \approx 0.82 \, \text{m}$$

In the fresh water the barge rides 2 cm lower than in the seawater.

Assess: The answer is precisely what we expected: The barge rides a bit (2 cm) lower because the fresh water is less dense than the seawater.

In fact, a shortcut would be to see that seawater is 3% more dense than fresh water, so the ship will ride 3% deeper in the fresh water. This gives exactly the same answer: $0.80 \text{ m} \times 103\% = 0.0824 \text{ m}$.

P13.25. Prepare: The buoyant force on the aluminum block is given by Archimedes' principle. The density of aluminum and ethyl alcohol are $\rho_{Al} = 2700 \text{ kg/m}^3$ and $\rho_{ethyl\ alcohol} = 790 \text{ kg/m}^3$. The buoyant force F_B and the tension due to the string act vertically up, and the weight of the aluminum block acts vertically down. The block is submerged, so the volume of displaced fluid equals V_{Al}, the volume of the block.

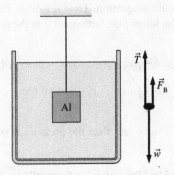

Solve: The aluminum block is in static equilibrium, so
$$\Sigma F_y = F_B + T - w = 0 \text{ N} \Rightarrow \rho_f V_{Al} g + T - \rho_{Al} V_{Al} g = 0 \text{ N} \Rightarrow T = V_{Al} g (\rho_{Al} - \rho_f)$$
$$T = (100 \times 10^{-6} \text{ m}^3)(9.80 \text{ m/s}^2)(2700 \text{ kg/m}^3 - 790 \text{ kg/m}^3) = 1.9 \text{ N}$$

Assess: The weight of the aluminum block is $\rho_{Al} V_{Al} g = 2.7 \text{ N}$. A similar order of magnitude for T is reasonable.

P13.27. Prepare: The buoyant force on the steel block is given by Archimedes' principle.

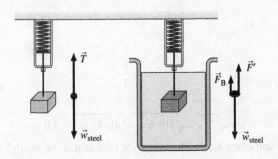

Solve: **(a)** In air the force on the spring is the weight of the block
$$F_{spring\ on\ block} = w = mg = \rho_{steel} V_{block} g = (7900 \text{ kg/m}^3)(0.10 \text{ m})^3 (9.8 \text{ m/s}^2) = 77.42 \text{ N} \approx 77 \text{ N}$$
to two significant figures.
(b) In oil
$$F_{spring\ on\ block} + F_B = 77.42 \text{ N}$$
$$\Rightarrow F_{spring\ on\ block} = 77.42 \text{ N} - \rho_{oil} V_{block} g = 77.42 \text{ N} - (900 \text{ kg/m}^3)(0.10 \text{ m})^3 (9.8 \text{ m/s}^2)$$
$$= 68.6 \text{ N} \approx 69 \text{ N}$$

Assess: The difference of $(77 \text{ N} - 69 \text{ N}) = 8 \text{ N}$ is due to the oil's buoyant force.

P13.31. Prepare: Treat the water as an ideal fluid. Two separate rivers merge to form one river. We will use the definition of the flow rate as given by Equation 13.13, $1\ L = 10^{-3}\ m^3$.

Solve: The volume flow rate in the Bernoulli River is the sum of the volume flow rate of River Pascal and River Archimedes.

$$Q_B = 5.0 \times 10^5\ L/s + 10.0 \times 10^5\ L/s = 15.0 \times 10^5\ L/s = 1500\ m^3/s$$

The flowrate is related to the fluid speed and the cross-sectional area by $Q = vA$. Thus

$$v_B = \frac{Q_B}{A_B} = \frac{1500\ m^3/s}{150\ m \times 10\ m^2} = 1.0\ m/s$$

Assess: A flow rate of 1.0 m/s is reasonable.

P13.35. Prepare: Please refer to Bernoulli's equation. Treat the oil as an ideal fluid obeying Bernoulli's equation. Consider the path connecting point 1 in the lower pipe with point 2 in the upper pipe a streamline.

Solve: Bernoulli's equation is

$$p_2 + \frac{1}{2}\rho v_2^2 + \rho g y_2 = p_1 + \frac{1}{2}\rho v_1^2 + \rho g y_1 \Rightarrow p_2 = p_1 + \frac{1}{2}\rho(v_1^2 - v_2^2) + \rho g(y_1 - y_2)$$

Using $p_1 = 200\ kPa = 200 \times 10^5\ Pa$, $\rho = 900\ kg/m^3$, $y_2 - y_1 = 10.0\ m$ $v_1 = 20.0\ m/s$, and $v_2 = 3.0\ m/s$, we get $p_2 = 1.1 \times 10^5\ Pa = 110\ kPa$.

Assess: We expect the pressure at point 2 to be less than the pressure at point 1. If this were not the case, the fluid would not flow from point 1 to point 2.

P13.37. Prepare: Treat the water as an ideal fluid obeying Bernoulli's equation.

$$\rho g(y_2 - y_1) = (p_1 - p_2) + \frac{1}{2}\rho(v_1^2 - v_2^2)$$

Consider a streamline connecting a point at the surface with a point in the hole. The pressure at the two points is the same, so $p_1 - p_2 = 0$. Further assumptions are that the area of the trough is so large that (1) it doesn't matter, and (2) the speed of the water at the top is zero ($v_1 = 0$).

Call $y_2 - y_1 = h = 0.45\ m$.

Solve: Since the pressures are equal we have

$$\rho g h = \frac{1}{2}\rho(v_1^2 - v_2^2)$$

Now set $v_1 = 0$ and cancel ρ.

$$gh = \frac{1}{2}v_2^2$$

Solve for v_2.

$$v_2 = \sqrt{2gh} = \sqrt{2(9.8\ m/s^2)(0.45\ m)} = 3.0\ m/s$$

Assess: The result is independent of the area of the trough, as long as it is big enough that we can assume $v_1 = 0$. The result $v_2 = \sqrt{2gh}$ is known as Torricelli's theorem.

P13.41. Prepare: For a cubic structure, volume is equal to the cube of the side. Volume is also equal to mass divided by density.

Solve: The volume of 0.197 kg of gold is

$$V = L^3 = \frac{M}{\rho_{Au}} \Rightarrow L = \left(\frac{(0.197\ kg)}{19\,300\ kg/m^3} \right)^{1/3} = 2.17\ cm$$

Assess: Large density for gold means that nearly half a pound of gold has a size of $\approx 2.2\ cm \times 2.2\ cm \times 2.2\ cm$.

P13.43. Prepare: $N = (M/M_A)N_A$, where N_A is Avogadro's number. Because the atomic mass number of Al is 27, one mole of Al has a mass of $M_A = 27\ g$.

Solve: The volume of the aluminum cube $V = 8.0 \times 10^{-6}$ m^3 and its mass is
$$M = \rho V = (2700 \text{ kg/m}^3)(8.0 \times 10^{-6} \text{ m}^3) = 0.0216 \text{ kg} = 21.6 \text{ g}$$

One mole of aluminum (^{27}Al) has a mass of 27 g. The number of atoms is
$$N = \left(\frac{6.02 \times 10^{23} \text{ atoms}}{1 \text{ mol}} \right) \left(\frac{1 \text{ mol}}{27 \text{ g}} \right) (21.6 \text{ g}) = 4.8 \times 10^{23} \text{ atoms}$$

Assess: A number slightly smaller than Avagadro's number is expected since we have slighty less than a mole of aluminum.

P13.45. Prepare: The pressure at the bottom of the tank is due to the atmosphere, the oil, and the water.
Solve: The pressure at the bottom of the oil layer is
$$p_{\text{bottom}} = p_0 + \rho_{\text{oil}} g d_{\text{oil}} + \rho_{\text{water}} g d_{\text{water}}$$
Solving for the height of the oil and inserting values we obtain
$$d_{\text{oil}} = (p_{\text{bottom}} - p_0 - \rho_{\text{water}} g d_{\text{water}})/\rho_{\text{oil}} g = 27 \text{ cm}$$
Assess: Compared to the depth of water, this is a reasonable number.

P13.51. Prepare: The buoyant force on the sphere is given by Archimedes' principle. The sphere is in static equilibrium.

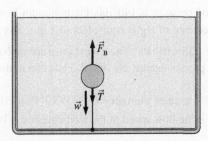

Solve: The free-body diagram on the sphere shows that
$$\Sigma F_y = F_B - T - w = 0 \text{ N} \Rightarrow F_B = T + w = \frac{1}{3}w + w = \frac{4}{3}w$$
$$\Rightarrow \rho_w V_{\text{sphere}} g = \frac{4}{3}\rho_{\text{sphere}} V_{\text{sphere}} g \Rightarrow \rho_{\text{sphere}} = \frac{3}{4}\rho_w = \frac{3}{4}(1000 \text{ kg/m}^3) = 750 \text{ kg/m}^3$$
Assess: We expected the sphere's density to be smaller than the water's because the sphere is tethered to the bottom.

P13.53. Prepare: The buoyant force on the rock is given by Archimedes' principle. We will use Newton's first law, as the rock is in equilibrium. A pictorial representation of the situation and the forces on the rock are shown.

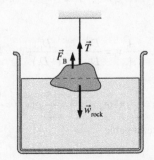

Solve: Because the rock is in static equilibrium, Newton's first law is

$$F_{net} = T + F_B - w_{rock} = 0 \text{ N} \Rightarrow T = \rho_{rock}V_{rock}g - \rho_{water}\left(\frac{1}{2}V_{rock}\right)g$$

$$= \left(\rho_{rock} - \frac{1}{2}\rho_{water}\right)V_{rock}g = \left(\rho_{rock} - \frac{1}{2}\rho_{water}\right)\left(\frac{m_{rock}g}{\rho_{rock}}\right) = \left(1 - \frac{\rho_{water}}{2\rho_{rock}}\right)m_{rock}g$$

Using $\rho_{rock} = 4800 \text{ kg/m}^3$ and $m_{rock} = 5.0 \text{ kg}$, we get $T = 44 \text{ N}$.

Assess: A buoyant force of $(5.0 \times 9.8 \text{ N} - 44 \text{ N}) \approx 5 \text{ N}$ is reasonable for this problem.

P13.55. Prepare: The pictorial representation that follows gives the relevant diameters of the syringe.

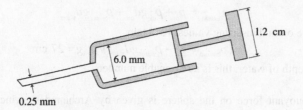

Solve: (a) Because the patient's blood pressure is 140/100, the minimum fluid pressure needs to be 100 mm of Hg above atmospheric pressure. Since 760 mm of Hg is equivalent to 1 atm and 1 atm is equivalent to 1.013×10^5 Pa, the minimum pressure is $100 \text{ mm} = 1.333 \times 10^4$ Pa. The excess pressure in the fluid is due to force F pushing on the internal 6.0-mm-diameter piston that presses against the liquid. Thus, the minimum force the nurse needs to apply to the syringe is

$$F = \text{fluid pressure} \times \text{area of plunger} = (1.333 \times 10^4 \text{ Pa})[\pi(0.003 \text{ m})^2] = 0.38 \text{ N}$$

(b) The flow rate is $Q = vA$, where v is the flow speed of the medicine and A is the cross-sectional area of the needle. Thus,

$$v = \frac{Q}{A} = \frac{2.0 \times 10^{-6} \text{ m}^3/2.0 \text{ s}}{\pi(0.125 \times 10^{-3} \text{ m})^2} = 20 \text{ m/s}$$

Assess: Note that the pressure in the fluid is due to F that is not dependent on the size of the plunger pad. Also note that the syringe is not drawn to scale.

P13.59. Prepare: Treat the water as an ideal fluid obeying Bernoulli's equation. Since the pipe is horizontal $y_1 = y_2$ so those terms drop out of the equation.

Solve:

$$p_1 + \frac{1}{2}\rho v_1^2 + \rho g y_1 = p_2 + \frac{1}{2}\rho v_2^2 + \rho g y_2 \Rightarrow \Delta p = p_1 - p_2 = \frac{1}{2}\rho(v_2^2 - v_1^2)$$

Use the continuity equation to find v_2.

$$v_2 = \frac{A_1}{A_2}v_1 = \frac{D_1^2}{D_2^2}v_1 = \frac{D_1^2}{(\frac{1}{2}D_1)^2}v_1 = 4v_1$$

Insert this into the previous equation.

$$\Delta p = \frac{1}{2}\rho(v_2^2 - v_1^2) = \frac{1}{2}\rho((4v_1)^2 - v_1^2) = \frac{1}{2}\rho(16v_1^2 - v_1^2) = \frac{15}{2}\rho(v_1^2) = \frac{15}{2}(1000 \text{ kg/m}^3)(1.4 \text{ m/s})^2 = 15 \text{ kPa}$$

Assess: This pressure difference is not unreasonable.

P13.63. Prepare: Treat the water as an ideal fluid obeying Bernoulli's equation. There is a streamline connecting point 1 in the wider pipe with point 2 in the narrower pipe.

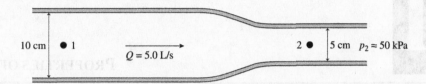

Solve: Bernoulli's equation, Equation 13.14, relates the pressure, water speed, and heights at points 1 and 2.

$$p_1 + \frac{1}{2}\rho v_1^2 + \rho g y_1 = p_2 + \frac{1}{2}\rho v_2^2 + \rho g y_2$$

For no height change $y_1 = y_2$. The flow rate is given as 5.0 L/s, which means

$$Q = v_1 A_1 = v_2 A_2 = 5.0 \times 10^{-3}\ \text{m}^3\text{/s} \Rightarrow v_1 = \frac{5.0 \times 10^{-3}\ \text{m}^3\text{/s}}{\pi (0.05\ \text{m})^2} = 0.6366\ \text{m/s}$$

$$\Rightarrow v_2 = v_1 \frac{A_1}{A_2} = (0.6366\ \text{m/s})\left(\frac{0.050\ \text{m}}{0.025\ \text{m}}\right)^2 = 2.546\ \text{m/s}$$

Bernoulli's equation now simplifies to

$$p_1 + \frac{1}{2}\rho v_1^2 = p_2 + \frac{1}{2}\rho v_2^2$$

$$\Rightarrow p_1 = p_2 + \frac{1}{2}\rho(v_2^2 - v_1^2) = 50 \times 10^3\ \text{Pa} + \frac{1}{2}(1000\ \text{kg/m}^3)[(2.546\ \text{m/s})^2 - (0.6366\ \text{m/s})^2] = 53\ \text{kPa}$$

Assess: Reducing the pipe size reduces the pressure because it makes $v_2 > v_1$.

P13.65. Prepare: We are considering a fluid with viscosity here, not an ideal fluid. We can see this from two hints in the problem. First, we are told that the water is at 20°C. For an ideal fluid, the temperature is irrelevant, but for a viscous fluid it is important because the viscosity depends on temperature. Second, we are told the lengths of the tubes, which is important for a viscous fluid but not for an ideal fluid. Thus we will assume a viscous fluid and use Poiseuille's equation.
Solve: According to Poiseuille's equation, the pressure drop along each section of the tube is

$$\Delta p = \frac{8\eta L Q}{\pi R^4}$$

Then the total pressure drop across the two sections of the tube is

$$\Delta p = \Delta p_1 + \Delta p_2 = \frac{8\eta L_1 Q_1}{\pi R_1^4} + \frac{8\eta L_2 Q_2}{\pi R_2^4} = \frac{8\eta L Q}{\pi}\left(\frac{1}{R_1^4} + \frac{1}{R_2^4}\right)$$

Here, we have used the fact that the two lengths L and the volume flow rates Q are the same in both sections. Thus

$$\Delta p = \frac{8(1.0 \times 10^{-3}\ \text{Pa} \cdot \text{s})(1.0\ \text{m})(0.02 \times 10^{-3}\ \text{m}^3\text{/s})}{\pi}\left(\frac{1}{(5.0 \times 10^{-4}\ \text{m})^4} + \frac{1}{(2.0 \times 10^{-3}\ \text{m})^4}\right)$$

$$= 8.2 \times 10^5\ \text{Pa}$$

This is the pressure difference between point P and the open end of the tube, which is at atmospheric pressure. Thus Δp is the difference between p_P and p_atmos, that is, it is the gauge pressure at P.

PART III

PROPERTIES OF MATTER

PptIII.7. Reason: If 25% of the 480 W is converted to mechanical energy of motion, then the other 75% is converted to thermal energy in his body. So $(0.75)(480 \text{ W}) = 360 \text{ W}$. The answer is B.

Assess: It takes some effort to stay cool on a strenuous bike ride.

PptIII.11. Reason: The buoyant force needs to be about the same magnitude as the weight force, so we compute the weight force on the balloon by assuming its mass is about the same as the mass of the displaced air. Read from the figure that the density of air at 10 km altitude is 0.4 kg/m³.

$$F_b = mg = \rho V g = (0.4 \text{ kg/m}^3)(12 \text{ m}^3)(9.8 \text{ m/s}^2) = 47 \text{ N} \approx 50 \text{ N}$$

The answer is A.

Assess: The weight of the balloon has hardly changed at all (being a little bit farther from the earth), so the buoyant force doesn't need to either.

PptIII.13. Reason: If the temperature were unchanged, then as the pressure halved the volume would double to 8.0 m³. But because the temperature in the balloon drops, this will somewhat shrink the gas volume, leading to a volume less than 8.0 m³. The answer is C.

Assess: We do not need to actually compute $2^{\frac{1}{1.4}}$ to know it is less than 2.

PptIII.17. Reason: The buoyancy force is in the opposite direction from the weight, so it is up. The drag force is in the opposite direction from the motion, so it is up too. The answer is A.

Assess: Since it is descending at a constant rate the sum of the three forces is zero.

PptIII.19. Reason: Estimate the area of the diaphragm to be $15 \text{ cm} \times 30 \text{ cm} = 0.045 \text{ m}^2$. Since pressure is force/area then

$$F = PA = (7.0 \text{ kPa})(0.045 \text{ m}^2) = 315 \text{ N} \approx 300 \text{ N}$$

Assess: The estimate is probably good to only one significant figure.

OSCILLATIONS

Q14.3. Reason: We are given the graph of x versus t. However, we want to think about the slope of this graph to answer velocity questions.

(a) When the x versus t graph is increasing, the particle is moving to the right. It has maximum speed when the positive slope of the x versus t graph is greatest. This occurs at 0 s, 4 s, and 8 s.

(b) When the x versus t graph is decreasing, the particle is moving to the left. It has maximum speed when the negative slope of the x versus t graph is greatest. This occurs at 2 s and 6 s.

(c) The particle is instantaneously at rest when the slope of the x versus t graph is zero. This occurs at 1 s, 3 s, 5 s, and 7 s.

Assess: This is reminiscent of material studied in Chapter 2; what is new is that the motion is oscillatory and the graph periodic.

Q14.5. Reason: Synthesis 14.1 shows that the maximum speed is proportional to the amplitude. For small angles doubling the angle corresponds to doubling the amplitude, so this increases her maximum speed by a factor of two.

Assess: The maximum acceleration also doubles.

Q14.7. Reason: The maximum kinetic energy is the same as the total mechanical energy. The total energy and amplitude of an oscillator are related by $E = kA^2/2$, we see that the energy is proportional to the square of the amplitude. If A is doubled, E will increase by a factor of four. That is $E_{New} = 4E_{Old} = 4(2\,J) = 8\,J$.

Assess: This question may also be answered using a more quantitative approach as outlined in Question 14.6.

Q14.11. Reason: The period of a block oscillating on a spring is given in Equation 14.26, $T = 2\pi\sqrt{m/k}$. We are told that $T_1 = 2.0$ s.

(a) In this case the mass is doubled, $m_2 = 2m_1$.

$$\frac{T_2}{T_1} = \frac{2\pi\sqrt{m_2/k}}{2\pi\sqrt{m_1/k}} = \sqrt{\frac{m_2}{m_1}} = \sqrt{\frac{2m_1}{m_1}} = \sqrt{2}$$

So $T_2 = \sqrt{2}T_1 = \sqrt{2}(2.0\,s) = 2.8$ s.

(b) In this case the spring constant is quadrupled, $k_2 = 2k_1$.

$$\frac{T_2}{T_1} = \frac{2\pi\sqrt{m/k_2}}{2\pi\sqrt{m/k_1}} = \sqrt{\frac{k_1}{k_2}} = \sqrt{\frac{k_1}{2k_1}} = \frac{1}{\sqrt{2}}$$

So $T_2 = \sqrt{2}T_1 = \sqrt{2}(2.0\,s) = 2.8$ s.

(c) The formula for the period does not contain the amplitude; that is, the period is independent of the amplitude. Changing (in particular, doubling) the amplitude does not affect the period, so the new period is still 2.0 s.

Assess: It is equally important to understand what *doesn't* appear in a formula. It is quite startling, really, the first time you realize it, that the amplitude doesn't affect the period. But this is crucial to the idea of simple harmonic motion. Of course, if the spring is stretched too far, out of its linear region, then the amplitude would matter.

Q14.13. Reason: If it behaves like a mass on a spring, then trimming the wings will reduce the mass, and this will increase the frequency because $f = \frac{1}{2\pi}\sqrt{\frac{k}{m}}$.

Assess: It would be easier to beat a wing quickly if it had less mass.

Q14.15. Reason: Reducing the mass increases the frequency because $f = \frac{1}{2\pi}\sqrt{\frac{k}{m}}$. So you would remove water.

Assess: Try it!

Q14.19. Reason: Natural frequency is the frequency that an oscillator will oscillate at on its own. You may drive an oscillator at a frequency other than its natural frequency.

Assess: If you are pushing a child in a swing, you build up the amplitude of the oscillation by driving the oscillator at its natural frequency. You can achieve resonance by driving an oscillator at its natural frequency.

Q14.23. Reason: (a) The unstretched equilibrium position is 20 cm. When we load the spring with 100 g we establish a new equilibrium position at 30 cm. When we pull the oscillator down to 40 cm (i.e., an additional 10 cm) and release it, it will oscillate with an amplitude of $A = 10$ cm. The correct choice is B.

(b) Knowing that 100 g stretched the spring 10 cm, we can determine the spring constant.

$$k = mg/x = (0.1\,\text{kg})(9.8\,\text{m/s}^2)/(0.10\,\text{m}) = 9.8\,\text{N/m}$$

Knowing the spring constant and the mass on the spring, we can determine the oscillation frequency as follows:

$$f = \sqrt{k/m}/2\pi = \sqrt{(9.8\,\text{N/m})/(0.10\,\text{kg})}/2\pi = 1.6\,\text{Hz}$$

The correct choice is C.

(c) The frequency of oscillation depends on k and m, neither of which would change on the moon. The correct choice is C.

Assess: These are rather small but acceptable values for the spring constant and frequency.

Q14.27. Prepare: Your arms act like simple pendulums and carrying the weights in your hands merely increases the mass of the bob, but that doesn't affect the frequency of oscillation, $f = \frac{1}{2\pi}\sqrt{\frac{g}{L}}$. The answer is B.

Assess: The mass of the bob doesn't appear in the equation for the frequency of a simple pendulum.

Q14.29. Reason: We see in Figure 14.26 that the cells on the basilar membrane close to the stapes correspond to higher frequencies.

The correct choice is B.

Assess: People often lose hearing sensitivity in the higher frequencies with age.

Problem

P14.3. Prepare: Your pulse or heartbeat is 75 beats per minute or 75 beats/60 s = 1.25 beats/s. The period is the inverse of the frequency, so we will use Equation 14.1.

Solve: The frequency of your heart's oscillations is

$$f = \frac{75\,\text{beats}}{60\,\text{s}} = 1.25\,\text{beats/s} = 1.3\,\text{Hz}$$

The period is the inverse of the frequency, hence

$$T = \frac{1}{f} = \frac{1}{1.3\,\text{Hz}} = 0.80\,\text{s}$$

Assess: A heartbeat of 1.3 beats per second means that one beat takes a little less than 1 second, which is what we obtained above.

P14.5. Prepare: For a small angle pendulum the restoring force is proportional to the displacement because $\sin \theta \approx \theta$.

Solve: If we model this as a Hooke's law situation, then doubling the distance will double the restoring force from 20 N to 40 N.

Assess: This would not work for angles much larger than $10°$.

P14.7. Model: The air-track glider attached to a spring is in simple harmonic motion. The glider is pulled to the right and released from rest at $t = 0$ s. It then oscillates with a period $T = 2.0$ s and a maximum speed $4v_{max} = 0$ cm/s $= 0.40$ m/s. While the amplitude of the oscillation can be obtained from Equation 14.13, the position of the glider can be obtained from Equation 14.10, $x(t) = A \cos(\frac{2\pi t}{T})$.

Solve: (a)

$$v_{max} = (2\pi A/T) \Rightarrow A = \frac{v_{max} T}{2\pi} = \frac{(0.40 \text{ m/s})(2.0 \text{ s})}{2\pi} = 0.127 \text{ m} = 0.13 \text{ m}$$

(b) The glider's position at $t = 0.25$ s is

$$x_{0.25 \text{ s}} = (0.127 \text{ m}) \cos\left[\frac{2\pi(0.25 \text{ s})}{2.0 \text{ s}}\right] = 0.090 \text{ m} = 9.0 \text{ cm}$$

Assess: At $t = 0.25$ s, which is less than one quarter of the time period, the object has not reached the equilibrium position and is still moving toward the left.

P14.11. Prepare: We will assume that ship and passengers are approximately in simple harmonic motion. Synthesis box 14.1 gives the maximum acceleration for an object in simple harmonic motion, $a_{max} = (2\pi f)^2 A$. $A = 1$ m and $f = 1/T = 1/15$ s $= 0.067$ Hz.

Solve: (a)

$$a_{max} = (2\pi f)^2 A = (2\pi \times 0.067 \text{ Hz})^2 (1 \text{ m}) = 0.2 \text{ m/s}^2$$

(b) To one significant figure, $g = 10 \text{ m/s}^2$, so the passenger's acceleration is about $\frac{1}{50} g$.

Assess: This is not a large acceleration, but it can play havoc with some people's stomachs.

P14.13. Prepare: The total side-to-side motion is $2A$. Solve $a_{max} = (2\pi f)^2 A$ for A. $a_{max} = (0.020)(9.80 \text{ m/s}^2)$.

Solve:

$$2A = 2\frac{a_{max}}{(2\pi f)^2} = 2\frac{(0.020)(9.80 \text{ m/s}^2)}{(2\pi(0.17 \text{ Hz}))^2} = 34 \text{ cm}$$

Assess: The height of the building is not needed.

P14.15. Prepare: Use the equation for v_{max}.

Solve: Solve the equation for A.

$$v_{max} = 2\pi f A \Rightarrow A = \frac{v_{max}}{2\pi f} = \frac{2.5 \text{ m/s}}{2\pi(250 \text{ Hz})} = 1.6 \text{ mm}$$

Assess: This seems about the right size for a bumblebee.

P14.19. Prepare: The block attached to the spring is in simple harmonic motion. The period of an oscillating mass on a spring is given by Equation 14.26.

Solve: The period of an object attached to a spring is

$$T = 2\pi\sqrt{\frac{m}{k}} = T_0 = 2.00 \text{ s}$$

where m is the mass and k is the spring constant.

(a) For mass $= 2m$,

$$T = 2\pi\sqrt{\frac{2m}{k}} = (\sqrt{2})T_0 = 2.83\,\text{s}$$

(b) For mass $\frac{1}{2}m$,

$$T = 2\pi\sqrt{\frac{\frac{1}{2}m}{k}} = T_0/\sqrt{2} = 1.41\,\text{s}$$

(c) The period is independent of amplitude. Thus $T = T_0 = 2.00\,\text{s}$.

(d) For a spring constant $= 2k$,

$$T = 2\pi\sqrt{\frac{m}{2k}} = T_0/\sqrt{2} = 1.41\,\text{s}$$

Assess: As would have been expected, increase in mass leads to slower simple harmonic motion.

P14.21. Prepare: The oscillating mass is in simple harmonic motion. The position of the oscillating mass is given by $x(t) = (2.0\,\text{cm})\cos(10t)$, where t is in seconds. We will compare this with Equation 14.10.

Solve: (a) The amplitude $A = 2.0\,\text{cm}$.

(b) The period is calculated as follows:

$$\frac{2\pi}{T} = 10\,\text{rad/s} \Rightarrow T = \frac{2\pi}{10\,\text{rad/s}} = 0.63\,\text{s}$$

(c) The spring constant is calculated from Equation 14.26 as follows:

$$\frac{2\pi}{T} = \sqrt{\frac{k}{m}} \Rightarrow k = m\left(\frac{2\pi}{T}\right)^2 = (0.050\,\text{kg})(10\,\text{rad/s})^2 = 5.0\,\text{N/m}$$

(d) The maximum speed from Equation 14.26 is

$$v_{\text{max}} = 2\pi f A = \left(\frac{2\pi}{T}\right)A = (10\,\text{rad/s})(2.0\,\text{cm}) = 20\,\text{cm/s}$$

(e) The total energy from Equation 14.21 is

$$E = \frac{1}{2}kA^2 = \frac{1}{2}(5.0\,\text{N/m})(0.02\,\text{m})^2 = 1.0\times10^{-3}\,\text{J}$$

(f) At $t = 0.40\,\text{s}$, the velocity from Equation 14.12 is

$$v_x = -(20.0\,\text{cm/s})\sin[(10\,\text{rad/s})(0.40\,\text{s})] = 15\,\text{cm/s}$$

Assess: Velocity at $t = 0.40\,\text{s}$, is less than the maximum velocity, as would be expected.

P14.25. Prepare: Assume a small angle of oscillation so there is simple harmonic motion. We will use Equation 14.27 for the pendulum's time period.

Solve: The period of the pendulum is

$$T_0 = 2\pi\sqrt{\frac{L_0}{g}} = 4.00\,\text{s}$$

(a) The period is independent of the mass and depends only on the length. Thus $T = T_0 = 4.00\,\text{s}$.

(b) For a new length $L = 2L_0$,

$$T = 2\pi\sqrt{\frac{2L_0}{g}} = \sqrt{2}T_0 = 5.66\,\text{s}$$

(c) For a new length $L = L_0/2$,

$$T = 2\pi\sqrt{\frac{L_0/2}{g}} = \frac{1}{\sqrt{2}}T_0 = 2.83\,\text{s}$$

(d) The period is independent of the amplitude as long as there is simple harmonic motion. Thus $T = 4.00\,\text{s}$.

P14.27. Prepare: Because the angle of displacement is less than $10°$, the small-angle approximation holds and the pendulum exhibits simple harmonic motion. We will use Equation 14.27 and $g = 9.80 \text{ m/s}^2$.

Solve: The period is $T = 12.0 \text{ s}/10 \text{ oscillations} = 1.20 \text{ s}$ and is given by the formula

$$T = 2\pi\sqrt{\frac{L}{g}} \Rightarrow L = \left(\frac{T}{2\pi}\right)^2 g = \left(\frac{1.20 \text{ s}}{2\pi}\right)^2 (9.80 \text{ m/s}) = 35.7 \text{ cm}$$

Assess: A length of 35.7 cm for the simple pendulum is reasonable.

P14.31. Prepare: Treating the lower leg as a physical pendulum we can determine the moment of inertia by combining $T = 2\pi\sqrt{I/mgL}$ and $T = 1/f$.

Solve: Combining the above expressions and solving for the moment of inertia we obtain

$$I = mgL/(2\pi f)^2 = (5.0 \text{ kg})(9.80 \text{ m/s}^2)(0.18 \text{ m})/[2\pi(1.6 \text{ Hz})]^2 = 8.7\times10^{-2} \text{ kg}\cdot\text{m}^2$$

Assess: NASA determines the moment of inertia of the shuttle in a similar manner. It is suspended from a heavy cable, allowed to oscillate about its vertical axis of symmetry with a very small amplitude, and from the period of oscillation one may determine the moment of inertia. This arrangement is called a torsion pendulum.

P14.35. Prepare: The motion is a damped oscillation. The maximum displacement or amplitude of the oscillation at time t is given by Equation 14.29, $x_{max}(t) = Ae^{-t/\tau}$, where τ is the time constant. Using $x_{max} = 0.368 A$ and $t = 10.0 \text{ s}$, we can calculate the time constant.

Solve: From Equation 14.29.

$$0.368A = Ae^{-10.0 \text{ s}/\tau} \Rightarrow \ln(0.368) = \frac{-10.0 \text{ s}}{\tau} \Rightarrow \tau = -\frac{10.0 \text{ s}}{\ln(0.368)} = 10.0 \text{ s}$$

Assess: The above result says that the oscillation decreases to about 37% of its initial value after one time constant.

P14.39. Prepare: Assume damped oscillations. We examine the graph carefully.

Solve: It looks from the graph that the period is 0.5 s so $f = 2 \text{ Hz}$. Looking at the peaks, it appears the amplitude has decreased to about 37% of its value after only 0.25 s, so that is our guess for τ.

Assess: You want the time constant for damping on your car to be short.

P14.41. Prepare: We will model the child on the swing as a simple small-angle pendulum. To make the amplitude grow large quickly we want to drive (push) the oscillator (child) at the natural resonance frequency. In other words, we want to wait the natural period between pushes.

Solve:

$$T = 2\pi\sqrt{\frac{L}{g}} = 2\pi\sqrt{\frac{2.0 \text{ m}}{9.8 \text{ m/s}^2}} = 2.8 \text{ s}$$

Assess: You could also increase the amplitude by pushing every other time (every $2T$), but that would not make the amplitude grow as quickly as pushing every period.

The mass of the child was not needed; the answer is independent of the mass.

P14.43. Prepare: Given the mass and the resonant frequency, we can determine the effective spring constant using the relationship $\omega = 2\pi f = \sqrt{k/m}$.

Solve: Solving the above expression for the spring constant, obtain
$$k = (2\pi f)^2 m = [2\pi(29 \text{ Hz})]^2 (7.5\times10^{-3} \text{ kg}) = 250 \text{ N/m}$$

Assess: As spring constants go, this is a fairly large value, however the musculature holding the eyeball in the socket is strong and hence will have a large effective spring constant.

P14.47. Prepare: The vertical oscillations constitute simple harmonic motion. A pictorial representation of the spring and the ball is shown in the following figure. The period and frequency of oscillations are

$$T = \frac{20\text{ s}}{30\text{ oscillations}} = 0.6667\text{ s} \quad \text{and} \quad f = \frac{1}{T} = \frac{1}{0.6667\text{ s}} = 1.50\text{ Hz}$$

Since k is known, we can obtain the mass m using Equation 14.26.

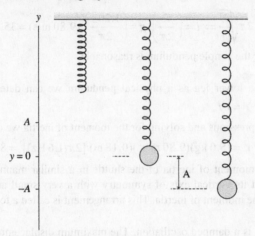

Solve: (a) The mass can be found as follows:

$$f = \frac{1}{2\pi}\sqrt{\frac{k}{m}} \Rightarrow m = \frac{k}{(2\pi f)^2} = \frac{15.0\text{ N/m}}{[2\pi(1.50\text{ Hz})]^2} = 0.169\text{ kg}$$

(b) The maximum speed is given by Equation 14.26, $v_{\max} = 2\pi f A = 2\pi(1.50\text{ Hz})(0.0600\text{ m}) = 0.565\text{ m/s}$.

Assess: Both the mass of the ball and its maximum speed are reasonable.

P14.49. Prepare: The vertical oscillations constitute simple harmonic motion. We will use Equation 14.26.

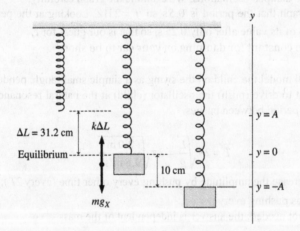

Solve: At the equilibrium position, the net force on mass m on Planet X is the following:

$$F_{\text{net}} = k\Delta L - mg_X = 0\text{ N} \Rightarrow \frac{k}{m} = \frac{g_X}{\Delta L}$$

For simple harmonic motion, Equation 14.26 yields $k/m = (2\pi f)^2$, thus

$$(2\pi f)^2 = \frac{g_X}{\Delta L} \Rightarrow g_X = \left(\frac{2\pi}{T}\right)^2 \Delta L = \left(\frac{2\pi}{14.5\text{ s}/10}\right)^2 (0.312\text{ m}) = 5.86\text{ m/s}^2$$

Assess: This value of g is of the same order of magnitude as the one for the earth, and would thus seem to be reasonable.

P14.51. Prepare: The vertical mass/spring systems are in simple harmonic motion.
Solve: (a) For system A, y is positive for one second as the mass moves downward and reaches maximum negative y after two seconds. It then moves upward and reaches the equilibrium position, $y = 0$, at $t = 3$ seconds. The maximum speed while traveling in the upward direction thus occurs at $t = 3.0$ s. The frequency of oscillation is 0.25 Hz.

(b) For system B, all the mechanical energy is potential energy when the position is at maximum amplitude, which for the first time is at $t = 1.5$ s. The time period of system B is thus 6.0 s.

(c) Spring/mass A undergoes three oscillations in 12 s, giving it a period $T_A = 4.0$ s. Spring/mass B undergoes two oscillations in 12 s, giving it a period $T_B = 6.0$ s. From Equation 14.26, we have

$$T_A = 2\pi\sqrt{\frac{m_A}{k_A}} \quad \text{and} \quad T_B = 2\pi\sqrt{\frac{m_B}{k_B}} \Rightarrow \frac{T_A}{T_B} = \sqrt{\left(\frac{m_A}{m_B}\right)\left(\frac{k_B}{k_A}\right)} = \frac{4.0\text{ s}}{6.0\text{ s}} = \frac{2}{3}$$

If $m_A = m_B$, then

$$\frac{k_B}{k_A} = \frac{4}{9} \Rightarrow \frac{k_A}{k_B} = \frac{9}{4} = 2.25$$

Assess: It is important to learn how to read a graph.

P14.55. Prepare: The compact car is in simple harmonic motion. The mass on each spring for the empty car is $(1200 \text{ kg})/4 = 300$ kg. However, the car carrying four persons means that each spring has, on the average, an additional mass of 70 kg. For both parts we will use Equation 14.27.
Solve: First calculate the spring constant as follows:

$$f = \frac{1}{2\pi}\sqrt{\frac{k}{m}} \Rightarrow k = m(2\pi f)^2 = (300 \text{ kg})[2\pi(2.0 \text{ Hz})]^2 = 4.74\times10^4 \text{ N/m}$$

Now reapply the equation with $m = 370$ kg, so

$$f = \frac{1}{2\pi}\sqrt{\frac{k}{m}} \Rightarrow \frac{1}{2\pi}\sqrt{\frac{4.74\times10^4 \text{ N/m}}{370 \text{ kg}}} = 1.8 \text{ Hz}$$

Assess: A small frequency change from the additional mass is reasonable because frequency is inversely proportional to the square root of the mass.

P14.59. Prepare: When the block is displaced from the equilibrium position one spring is compressed and exerts a restoring force on the block while the other spring is stretched and also exerts a restoring force on the block. These two forces have the same magnitude since the springs are identical and a stretch or compression of the same amount produces the same restoring force.
Solve: (a) Use Hooke's law. The restoring force is the sum of the restoring forces from the two springs.
$$F = -kx = -(20 \text{ N/m} + 20 \text{ N/m})(-0.010 \text{ m}) = 0.40 \text{ N}$$

(b) The restoring force would be the same if the springs were on the same side of the block. For springs in parallel like that, the total k_{tot} is the sum of the two k's.

$$k_{tot} = k_1 + k_2 = 20 \text{ N/m} + 20 \text{ N/m} = 40 \text{ N/m}$$

(c) f_{two} is the frequency for the system with two springs.

$$f_{two} = \frac{1}{2\pi}\sqrt{\frac{k_{tot}}{m}} = \frac{1}{2\pi}\sqrt{\frac{40 \text{ N/m}}{2.5 \text{ kg}}} = 0.64 \text{ Hz}$$

Assess: The answer seems reasonable.
If the two k's are the same $k_1 = k_2 = k$ (as in this case), you can see the general formula would be

$$f_{two} = \frac{1}{2\pi}\sqrt{\frac{k_1 + k_2}{m}} = \frac{1}{2\pi}\sqrt{\frac{2k}{m}} = \sqrt{2}\left(\frac{1}{2\pi}\sqrt{\frac{k}{m}}\right) = f_{one}$$

The frequency with two identical springs is $\sqrt{2}$ times the frequency with one spring.
An even more general result can be obtained with similar reasoning even where the two k's differ, $f_{two}^2 = f_1^2 + f_2^2$.

P14.63. Prepare: The maximum speed occurs at the equilibrium position.
Solve:

$$v_{max} = \frac{2\pi}{T} A = \frac{2\pi}{0.50 \text{ s}} (0.30 \text{ m}) = 0.377 \text{ m/s}$$

After it detaches it is in free fall.

$$H = \frac{v_0^2}{2g} = \frac{(3.77 \text{ m/s})^2}{2(9.8 \text{ m/s}^2)} = 0.7251 \text{ m} \approx 73 \text{ cm}$$

Assess: This is a safe and reasonable number.

P14.65. Prepare: The oscillator is in simple harmonic motion and is damped, so we will use Equation 14.29.
Solve: The maximum displacement, or amplitude, of a damped oscillator decreases as $x_{max}(t) = Ae^{-t/\tau}$, where τ is the time constant. We know $x_{max}/A = 0.60$ at $t = 50 \text{ s}$, so we can find τ as follows:

$$-\frac{t}{\tau} = \ln\left(\frac{x_{max}(t)}{A}\right) \Rightarrow \tau = \frac{50 \text{ s}}{\ln(0.60)} = 97.88 \text{ s}$$

Now we can find the time t_{30} at which $x_{max}/A = 0.30$

$$t_{30} = -\tau \ln\left(\frac{x_{max}(t)}{A}\right) = -(97.88 \text{ s})\ln(0.30) = 118 \text{ s}$$

The undamped oscillator has a frequency $f = 2 \text{ Hz} = 2$ oscillations per second. Then the number of oscillations before the amplitude decays to 30% of its initial amplitude is $N = f \cdot t_{30} = (2 \text{ oscillations/s}) \cdot (118 \text{ s}) = 236$ oscillations or 240 oscillations to two significant figures.

15

TRAVELING WAVES AND SOUND

Q15.1. Reason: (a) In a transverse wave, the thing or quantity that is oscillating, such as the particles in a string, oscillates in a direction that is transverse (perpendicular) to the direction of the propagation of the wave.
(b) Vibrations of a bass guitar string are a form of transverse wave. You can see that the oscillation is perpendicular to the string.
Assess: The plucking action makes the segment of the string move perpendicular to the string, but the disturbance is propagated along the string.

Q15.5. Reason: Equation 15.2, $v_{string} = \sqrt{T_s/\mu}$, gives the wave speed on a stretched string with tension T_s and linear mass density $\mu = m/L$. We will investigate how T_s and μ are changed in each case below and how that affects the wave speed. Use a subscript 1 for the original string, and a subscript 2 for the altered string.
We are given that $(v_{string})_1 = 200$ cm/s.

(a) $(T_s)_2 = 2(T_s)_1 \qquad \mu_2 = \mu_1$

$$\frac{(v_{string})_2}{(v_{string})_1} = \frac{\sqrt{(T_s)_2/\mu_2}}{\sqrt{(T_s)_1/\mu_1}} = \sqrt{\frac{(T_s)_2/\mu_2}{(T_s)_1/\mu_1}} = \sqrt{\frac{2(T_s)_1/\mu_1}{(T_s)_1/\mu_1}} = \sqrt{2}$$

$$(v_{string})_2 = \sqrt{2}(v_{string})_1 = \sqrt{2}(200 \text{ cm/s}) = 280 \text{ cm/s}$$

(b) $(T_s)_2 = (T_s)_1 \qquad m_2 = 4m_1 \Rightarrow \mu_2 = 4\mu_1$

$$\frac{(v_{string})_2}{(v_{string})_1} = \frac{\sqrt{(T_s)_2/\mu_2}}{\sqrt{(T_s)_1/\mu_1}} = \sqrt{\frac{(T_s)_2/\mu_2}{(T_s)_1/\mu_1}} = \sqrt{\frac{(T_s)_1/4\mu_1}{(T_s)_1/\mu_1}} = \frac{1}{2}$$

$$(v_{string})_2 = \frac{1}{2}(v_{string})_1 = \frac{1}{2}(200 \text{ cm/s}) = 100 \text{ cm/s}$$

(c) $(T_s)_2 = (T_s)_1 \qquad L_2 = 4L_1 \Rightarrow \mu_2 = \frac{1}{4}\mu_1$

$$\frac{(v_{string})_2}{(v_{string})_1} = \frac{\sqrt{(T_s)_2/\mu_2}}{\sqrt{(T_s)_1/\mu_1}} = \sqrt{\frac{(T_s)_2/\mu_2}{(T_s)_1/\mu_1}} = \sqrt{\frac{(T_s)_1/\frac{1}{4}\mu_1}{(T_s)_1/\mu_1}} = 2$$

$$(v_{string})_2 = 2(v_{string})_1 = 2(200 \text{ cm/s}) = 400 \text{ cm/s}$$

(d) $(T_s)_2 = (T_s)_1 \qquad m_2 = 4m_1 \quad$ and $\quad L_2 = 4L_1 \Rightarrow \mu_2 = \mu_1$

$$\frac{(v_{string})_2}{(v_{string})_1} = \frac{\sqrt{(T_s)_2/\mu_2}}{\sqrt{(T_s)_1/\mu_1}} = \sqrt{\frac{(T_s)_2/\mu_2}{(T_s)_1/\mu_1}} = \sqrt{\frac{(T_s)_1/\mu_1}{(T_s)_1/\mu_1}} = 1$$

$$(v_{string})_2 = (v_{string})_1 = 200 \text{ cm/s}$$

Assess: Notice as in part **(d)** that if both the mass and length of the string are increased by the same factor, then μ is not changed, so the speed is the same (with no change in tension).

Q15.7. Reason: The speed of sound in air depends on the temperature of the air. The distance across the stadium can be measured to give the path length of the sound. Then, you can take the path length and divide by the time between the emission and detection of the pulse to get the speed of the sound. Finally, once the speed of sound is known, you can find the corresponding temperature either by consulting a chart like Table 15.1 or by using Equation 15.3.

Assess: The advantage of measuring temperature this way is that it gives you an idea of the average temperature for the whole stadium. It also determines the temperature quickly—the time of measurement is simply the time it takes for sound to travel across the stadium.

Q15.9. Reason: Since the wave is traveling to the left, the snapshot has the same shape as the history graph. To understand why, consider that the history graph tells us about the displacement at one point in space. As the wave moves to the left, that point witnesses spots on the wave further to the right. Thus increasing t by Δt on the history graph has the same effect as increasing x by $v\Delta t$ on the snapshot graph, where v is the speed of the wave. Consequently the history graph and snapshot graph have the same shape. The only difference is that in going from the history graph to the snapshot graph, the horizontal axis is scaled by a factor of v.

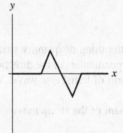

Assess: A similar argument can be used to show that if a wave is traveling to the *right*, the snapshot graph can be obtained from the history graph by reflecting the graph about the y-axis and scaling the horizontal axis by a factor of v.

Q15.13. Reason: The bat's forward motion means the reflected pulse will be Doppler shifted to a higher frequency; consequently, the bat should decrease the frequency of the emitted pulse so the higher reflected pulse will be in the correct frequency range.

Assess: If the bat increases its forward speed, then it should further decrease the emitted frequency.

Q15.17. Reason: Because the bullet is traveling faster than the speed of sound it creates a shock wave—a little sonic boom. This is what people at a distance from the gun may hear louder.

Assess: Even the sharp crack of a bull whip is a sonic boom, as the tip of the whip is traveling faster than the speed of sound in air.

Q15.19. Reason: The guns are calibrated for motion directly toward or away from the gun. If a car is moving perpendicular to the line of sight of the gun then it has no motion in the line of sight and there will be no Doppler shift. If the car is coming or going at some other angle then there will be a Doppler shift, but not as big as if the car were traveling along the line of sight of the gun at the same speed. So, if the radar gun reports that a car is moving at a certain speed, it is moving at least that fast, and faster if the motion isn't along the line of sight.

Assess: The police must calibrate their radar guns periodically, but if they have done so and it shows you were speeding, don't try to argue out of it by saying you were going at an angle to the line of sight.

Q15.21. Reason: The fundamental relationship for periodic waves is given in Equation 15.10, $v = \lambda f$.

We are asked for the frequency of an electromagnetic wave (which travels at a speed of 3.00×10^8 m/s) of a given wavelength ($\lambda = 400$ nm).

$$f = \frac{v}{\lambda} = \frac{3.00\times10^8 \text{ m/s}}{400\times10^{-9} \text{ m}} = 7.50\times10^{14} \text{ Hz}$$

The correct choice is D.

Assess: Note the tremendously high frequency of visible light: seven and a half hundred trillion hertz! You may not remember the exact number (which, of course, varies for different colors), but it would be fun to tuck away the tidbit that $f \approx 10^{14}$ Hz for visible light in your brain for later recall.

This calculation is easily done in one's head (keep track of the significant figures).

$$f = \frac{v}{\lambda} = \frac{3.00 \times 10^8 \text{ m/s}}{400 \times 10^{-9} \text{ m}} = \frac{3}{4}\left(\frac{10^8}{10^{-7}}\right) \text{Hz} = 0.75 \times 10^{15} \text{ Hz} = 7.50 \times 10^{14} \text{ Hz}$$

Q15.23. Reason: The fundamental relationship for periodic waves is given in Equation 15.10, $v = \lambda f$. We are told that $\lambda = 32$ cm. We'll compute the frequency from $f = 1/T = 1/0.20 \text{ s} = 5.0 \text{ Hz}$, $v = \lambda f = (32 \text{ cm})$ (5.0 Hz) = 160 cm/s.
The correct choice is D.
Assess: Equation 15.9 is even more directly applicable, but the customary form (that you should remember) is Equation 15.10. The amplitude was unneeded information.

Q15.25. Reason: When the measured frequency is lower than f_s it is because the source is moving away from you. This is the case from $t = 0$ s to $t = 2$ s. After that the measured frequency is higher than f_s so the source is moving toward you.
The correct choice is D.
Assess: In a doppler shift situation the measured frequency is constant as long as the source's velocity toward you (or away from you) is constant. The measured frequency doesn't keep increasing as the source gets closer and closer.

Problems

P15.3. Prepare: Assume the speed of sound is the 20°C value 343 m/s. A pictorial representation of the problem is given.

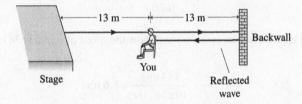

Solve: The time elapsed from the stage to you in the middle row is the distance divided by the speed of sound.

$$\frac{13 \text{ m}}{343 \text{ m/s}} = 0.0379 \text{ s}$$

The time elapsed for the same sound to get reflected from the back wall and reach your ear is

$$\frac{26 \text{ m} + 13 \text{ m}}{343 \text{ m/s}} = 0.114 \text{ s}$$

The difference in the two times is thus $0.114 \text{ s} - 0.038 \text{ s} = 0.076 \text{ s}$.
Assess: Due to the relatively large speed of sound, the observed small time difference would be expected.

P15.5. Prepare: We will assume the wave speed in water to be 1480 m/s, as given in Table 15.1.
Solve: The time interval for the sound wave to travel a distance of 3200 km through water is

$$\Delta t = \frac{\Delta x}{v_{\text{water}}} = \frac{3200 \text{ km}}{1480 \text{ m/s}} = 2162 \text{ s} = 36 \text{ min}$$

Assess: A distance of 3200 km is quite large when the speed of sound in water is about 1.5 km/s, so a time of 36 min is reasonable.

P15.7. Prepare: We will compute the time it takes for each of the waves to travel 45 km and then subtract to get the difference.

Solve:

$$\Delta t_P = \frac{\Delta x}{v_P} = \frac{45 \text{ km}}{5000 \text{ m/s}} = 9.0 \text{ s}$$

$$\Delta t_S = \frac{\Delta x}{v_S} = \frac{45 \text{ km}}{3000 \text{ m/s}} = 15.0 \text{ s}$$

The difference in arrival times between the P and S waves is 15.0 s – 9.0 s = 6.0 s.
Assess: This seems to be on the order of how quickly things happen in earthquakes.

P15.11. Prepare: This is a wave traveling at constant speed to the right at 1 m/s.
Solve: This is the history graph of the wave at $x = 0$ m. The graph shows that the $x = 0$ m point of the medium first sees the negative portion of the pulse wave at $t = 1$ s. Thus, the snapshot graph of this wave at $t = 1$ s must have the leading negative portion of the wave at $x = 0$ m.

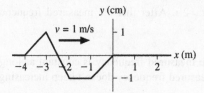

Snapshot graph at $t = 1.0$ s

P15.13. Prepare: We will use Equation 15.9 to find the wave speed.
Solve: The wave speed is

$$v = \frac{\lambda}{T} = \frac{2.0 \text{ m}}{0.20 \text{ s}} = 10 \text{ m/s}$$

P15.15. Prepare: Since the lab temperature is 20°C, we know the speed of sound is 343 m/s, as given in Table 15.1.
Solve: (a) $f = 40$ kHz

$$\lambda = \frac{v}{f} = \frac{343 \text{ m/s}}{40 \times 10^3 \text{ Hz}} = 8.6 \text{ mm}$$

(b) For the round trip the distance is 5.0 m.

$$\Delta t = \frac{\Delta x}{v_x} = \frac{5.0 \text{ m}}{343 \text{ m/s}} = 0.015 \text{ s} = 15 \text{ ms}$$

Assess: 15 ms is quick by human reaction time standards, but it is easy to have electronics in the detector record time intervals such as this.

P15.19. Prepare: This is a history graph.
Solve: The amplitude of the wave is the maximum displacement which is 6.0 cm.
The period of the wave is 0.60 s, so the frequency $f = 1/T = 1/0.60 \text{ s} = 1.667 \text{ Hz}$, or 1.7 Hz to two significant figures.
The wavelength, using Equation 15.10, is

$$\lambda = \frac{v}{f} = \frac{2 \text{ m/s}}{1.667 \text{ Hz}} = 1.2 \text{ m}$$

Assess: It is important to know how to read information from graphs.

P15.23. Prepare: We will use the fundamental relation for periodic waves in Equation 15.10 to solve for the wavelength. We must first look up the speed of sound in water; Table 15.1 says it is 1480 m/s.
Solve:

$$\lambda = \frac{v}{f} = \frac{1480 \text{ m/s}}{100 \times 10^3 \text{ Hz}} = 15 \text{ mm} = 1.5 \text{ cm}$$

Assess: Because the speed of sound in water is over four times the speed of sound in air, dolphins must be quick to process the sonar information.

P15.27. Prepare: Microwaves are electromagnetic waves that travel with a speed of 3×10^8 m/s.

Solve: (a) The frequency of the microwave is

$$f_{\text{microwaves}} = \frac{c}{\lambda} = \frac{3.0 \times 10^8 \text{ m/s}}{3.0 \times 10^{-2} \text{ m}} = 1.0 \times 10^{10} \text{ Hz} = 10 \text{ GHz}$$

(b) The time for the microwave signal to travel is

$$t = \frac{50 \text{ km}}{v_{\text{air}}} = \frac{50 \times 10^3 \text{ m}}{3.0 \times 10^8 \text{ m}} = 0.17 \text{ ms}$$

Assess: A small time of 0.17 ms for the microwaves to cover a distance of 50 km shows that the electromagnetic waves travel very fast.

P15.29. Prepare: The power (or energy/time) is the intensity multiplied by the area. The intensity is $I = 1.0 \times 10^{-6}$ W/m². We can deduce from the information given that the area of the eardrum is $A = \pi R^2 = \pi (4.2 \text{ mm})^2 = 5.54 \times 10^{-5}$ m².

Solve:

$$P = Ia = (1.0 \times 10^{-6} \text{ W/m}^2)(5.54 \times 10^{-5} \text{ m}^2) = 5.5 \times 10^{-11} \text{ W}$$

The energy delivered to your eardrum each second is 55 pJ.

Assess: This is an incredibly tiny amount of energy per second; you should be impressed that your ear can detect such small signals! Your eardrum moves back and forth about the width of 100 atoms in such cases!

P15.31. Prepare: We are asked to find the energy received by your back of area 30 cm × 50 cm in 1.0 h if the electromagnetic wave intensity is 1.4×10^3 W/m². The energy delivered to your back in time t is $E = Pt$, where P is the power of the electromagnetic wave. The intensity of the wave is $I = P/a$ where a is the area of your back.

Solve: The energy received by your back is

$$E = Pt = Iat = (0.80)(1400 \text{ W/m}^2)(0.30 \times 0.50 \text{ m}^2)(3600 \text{ s}) = 6.0 \times 10^5 \text{ J}$$

Assess: This is equivalent to receiving approximately 170 J of energy per second by your back. This energy is relatively large and will certainly lead to tanning.

P15.33. Prepare: If the solar panel produces electrical power with efficiency ε, then the power received from the sun is $P_E = (1.0 \text{ kW})/\varepsilon$. This is because if we multiply the power from the sun by ε, we should get $\varepsilon P_E = 1.0$ kW. If the area of the solar panel is a, then the intensity of radiation from the sun is $(1.0 \text{ kW})/\varepsilon a$. To find the intensity at some other distance from the sun, we use Equation 15.13, $I_1/I_2 = r_2^2/r_1^2$.

Sun

Earth

Saturn

Known
$r_S/r_E = 9.5$
$\varepsilon P_E = 1.0$ kW

Find
εP_S

Solve: Using the previous formula, we can find the intensity of sunlight at the location of Saturn's orbit.

$$I_S = I_E \left(\frac{r_E}{r_S}\right)^2 = \left(\frac{1.0 \text{ kW}}{\varepsilon a}\right)\left(\frac{1}{9.5}\right)^2 = \frac{11 \, W}{\varepsilon a}$$

Then using $P = Ia$, we find that $P_S = (11 \, W)/\varepsilon$, that is the solar panel would receive $(11 \, W)/\varepsilon$, of power from the sun. Finally, since the panels work with efficiency ε, the power which would be produced if the spacecraft were in orbit around Saturn is $\varepsilon P_S = 11$ W.

Assess: The reduction from 1000 W to 11 W corresponds to a factor of about 100. This is reasonable since intensity is inversely proportional to distance squared. At Saturn's orbit, the satellite is about 10 times farther away from the sun so the intensity of light it receives is reduced by a factor of about 100.

P15.37. Prepare: We will use Equation 15.12 and the definition of power.
Solve: The peak power of the light pulse is

$$P_{\text{peak}} = \frac{\Delta E}{\Delta t} = \frac{500 \text{ mJ}}{10 \text{ ns}} = \frac{0.500 \text{ J}}{1.0 \times 10^{-8} \text{ s}} = 5.0 \times 10^7 \text{ W}$$

The intensity during a pulse is

$$I_{\text{laser}} = \frac{P}{a} = \frac{5.0 \times 10^7 \text{ W}}{\pi (5.0 \ \mu\text{m})^2} = \frac{5.0 \times 10^7 \text{ W}}{7.85 \times 10^{-11} \text{ m}^2} = 6.4 \times 10^{17} \text{ W/m}^2$$

Assess: The laser produces light whose *intensity* is more than 14 orders of magnitude larger than the sun's intensity on the target.

P15.41. Solve: (a) If a source of spherical waves radiates uniformly in all directions, the ratio of the intensities at distances r_1 and r_2 is

$$\frac{I_1}{I_2} = \frac{r_2^2}{r_1^2} \Rightarrow \frac{I_{50 \text{ m}}}{I_{2 \text{ m}}} = \left(\frac{2 \text{ m}}{50 \text{ m}}\right)^2 = 1.6 \times 10^{-3} \Rightarrow I_{50 \text{ m}} = I_{2 \text{ m}}(1.6 \times 10^{-3}) = (2.0 \text{ W/m}^2)(1.6 \times 10^{-3}) = 3.2 \times 10^{-3} \text{ W/m}^2$$

(b) The sound intensity level is given from Equation 15.14.

$$\beta = (10 \text{ dB}) \log_{10}\left(\frac{I}{I_0}\right) = (10 \text{ dB}) \log_{10}\left(\frac{3.2 \times 10^{-3} \text{ W/m}^2}{1.0 \times 10^{-12} \text{ W/m}^2}\right) = 95 \text{ dB}$$

Assess: The power generated by the sound source is $P = I_{2 \text{ m}} [4\pi(2 \text{ m})^2] = (2.0 \text{ W/m}^2)(50.27) = 101$ W. This is a significant amount of power.

P15.43. Prepare: Knowing the sound intensity level $\beta_1 = 120$ dB at the point $r_1 = 5$ m, we can determine the intensity I_1 at this point. Since the power output of the speaker is a constant, knowing the intensity I_1 and distance r_1 from the speaker at one point, we can determine the intensity I_2 at a second point r_2. Knowing the intensity of the sound I_2 at this second point r_2, we can determine the sound intensity level β_2 at the second point.

Solve: The sound intensity may be determined from the sound intensity level as follows: $\beta_1 = (10 \text{ dB}) \log_{10} (I_1/I_0)$.

Inserting numbers: $120 \text{ dB} = (10 \text{ dB}) \log_{10} [I_1/(10^{-12} \text{ W/m}^2)]$ or $12 = \log_{10} [I_1/(10^{-12} \text{ W/m}^2)]$

Taking the antilog of both sides obtains $10^{12} = I_1/(10^{-12} \text{ W/m}^2)$ which may be solved for I_1 to obtain $I_1 = (10^{-12} \text{ W/m}^2)(10^{12}) = 1 \text{ W/m}^2$.

Knowing how the sound intensity is related to the power and that the power does not change, we can determine the sound intensity at any other point.

Inserting $a = 4\pi r^2$, into $P = I_1 a_1 = I_2 a_2$ obtain $I_2 = I_1(r_1/r_2)^2 = (1 \text{ W/m}^2)(5/35)^2 = 2.04 \times 10^{-2} \text{ W/m}^2$.

Finally, knowing the sound intensity at the second point r_2, we can determine the sound intensity level at this point by

$$\beta_2 = (10 \text{ dB}) \log_{10} (I_2/I_o) = (10 \text{ dB}) \log_{10} (2.04 \times 10^{-2}/1 \times 10^{-12}) = (10 \text{ dB}) \log_{10} (2.04 \times 10^{10})$$

continuing

$$\beta_2 = 10 \text{ dB}[\log_{10} (2.04) + \log_{10}(10^{10})] = 10 \text{ dB}[0.31 + 10] = 103 \text{ dB} \approx 100 \text{ dB}$$

Assess: This is still as loud as a pneumatic hammer. Either take ear protection or move farther away from the speaker.

P15.47. Prepare: The frequency of the opera singer's note is altered by the Doppler effect. The frequency is f_+ as the car approaches and f_- as it moves away. f_s is the frequency of the source. The speed of sound in air is 343 m/s.
Solve: **(a)** Using 90 km/hr = 25 m/s, the frequency as her convertible approaches the stationary person is

$$f_+ = \frac{f_s}{1 - v_s/v} = \frac{600 \text{ Hz}}{1 - \dfrac{25 \text{ m/s}}{343 \text{ m/s}}} = 650 \text{ Hz}$$

(b) The frequency as her convertible recedes from the stationary person is

$$f_- = \frac{f_s}{1 + v_s/v} = \frac{600 \text{ Hz}}{1 + \dfrac{25 \text{ m/s}}{343 \text{ m/s}}} = 560 \text{ Hz}$$

Assess: As would have been expected, the pitch is higher in front of the source than it is behind the source.

P15.49. Prepare: Sound frequency is altered by the Doppler effect. The frequency increases for an observer approaching the source and decreases for an observer receding from the source. You need to ride your bicycle away from your friend to lower the frequency of the whistle. We will use Equation 15.17.
Solve: The minimum speed you need to travel is calculated as follows:

$$f_- = \left(1 - \frac{v_s}{v}\right) f_s \Rightarrow 20 \text{ kHz} = \left(1 - \frac{v_s}{343 \text{ m/s}}\right)(21 \text{ kHz}) \Rightarrow v_s = 16 \text{ m/s}$$

Assess: A speed of 16 m/s corresponds to approximately 35 mph. This is a possible but very fast speed on a bicycle.

P15.53. Prepare: The frequency shift of the ultrasound reflected from blood moving in the artery may be determined by $\Delta f = \pm 2 f_s (v_s/v)$. Here f_s is the frequency of the ultra sound, v is the speed of the ultrasound in human tissue (1540 m/s) and v_s is the speed of the blood cell reflecting the ultrasound.
Solve:

$$\Delta f = \pm 2 f_s (v_s/v) = \pm 2(5.0 \times 10^6 \text{ Hz})[(0.20 \text{ m/s})/(1540 \text{ m/s})] = 1.3 \text{ kHz}$$

Assess: Compared to examples in the text, this is a reasonable number.

P15.55. Prepare: The explosive's sound travels down the lake and into the granite, and then it is reflected by the oil surface. A pictorial representation of the problem along with the speed of sound in water and granite is shown.

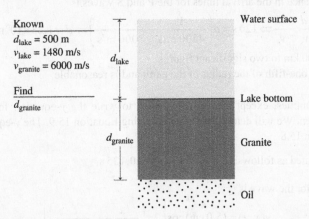

Solve: The echo time is equal to

$$t_{\text{echo}} = t_{\text{water down}} + t_{\text{granite down}} + t_{\text{granite up}} + t_{\text{water up}}$$

$$0.94 \text{ s} = \frac{500 \text{ m}}{1480 \text{ m/s}} + \frac{d_{\text{granite}}}{6000 \text{ m/s}} + \frac{d_{\text{granite}}}{6000 \text{ m/s}} + \frac{500 \text{ m}}{1480 \text{ m/s}} \Rightarrow d_{\text{granite}} = 790 \text{ m}$$

Assess: Drilling into granite for 3/4 km is not unreasonable.

P15.57. Prepare: To get the time needed for a wave to travel down a string, we need the length of the string and the speed of the wave. The speed comes from Equation 15.2. We are given the mass and length of the string from which we can find the mass density.

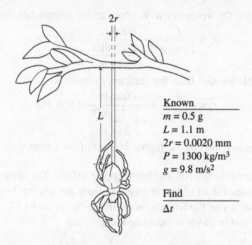

Known
m = 0.5 g
L = 1.1 m
2r = 0.0020 mm
P = 1300 kg/m³
g = 9.8 m/s²

Find
Δt

Solve: Since the silk is cylindrical, its volume is given by $V = \pi r^2 L$. From this equation we can find the linear density as follows:

$$\mu = \frac{m}{L} = \frac{\pi r^2 m}{\pi r^2 L} = \frac{\pi r^2 m}{V} = \pi r^2 \rho = \pi (1.0 \times 10^{-6}\ \text{m})^2 (1300\ \text{kg/m}^3) = 4.08 \times 10^{-9}\ \text{kg/m}$$

The speed of the wave on the string is $v = \sqrt{\dfrac{T}{\mu}} = \sqrt{\dfrac{4.9 \times 10^{-3}\ \text{N}}{4.08 \times 10^{-9}\ \text{kg/m}}} = 1.10 \times 10^3\ \text{m/s}$. The time needed for a pulse to

travel the length of the string is just the length of the string divided by the speed of waves on the string.

$$\Delta t = \frac{L}{v} = \frac{1.1\ \text{m}}{1.10 \times 10^3\ \text{m/s}} = 1.0\ \text{ms}$$

Assess: It is good for the spider that very little time is needed for it to receive this information. Notice that the speed of the wave is greater than the speed of sound in air.

P15.61. Solve: The difference in the arrival times for the P and S waves is

$$\Delta t = t_S - t_P = \frac{d}{v_S} - \frac{d}{v_P} \Rightarrow 120\ \text{s} = d\left(\frac{1}{4500\ \text{m/s}} - \frac{1}{8000\ \text{m/s}}\right) \Rightarrow d = 1.23 \times 10^6\ \text{m} = 1230\ \text{km}$$

This will be reported as 1200 km to two significant figures.
Assess: d is approximately one-fifth of the radius of the earth and is reasonable.

P15.65. Prepare: All quantities, except the period, needed to write the y-equation for a wave traveling in the negative x-direction are given. We will determine the period using Equation 15.9. The y-equation that we are asked to write will look like Equation 15.8.

Solve: The period is calculated as follows: $T = \dfrac{\lambda}{v} = \dfrac{0.50\ \text{m}}{4.0\ \text{m/s}} = 0.125\ \text{s}$.

The displacement equation for the wave is

$$y(x,\ t) = (5.0\ \text{cm})\cos\left[2\pi\left(\frac{x}{50\ \text{cm}} + \frac{t}{0.125\ \text{s}}\right)\right]$$

Assess: The positive sign in the cosine function's argument indicates motion along the $-x$ direction.

P15.67. Prepare: The function of x and t with the minus sign will be a wave traveling to the right.

Solve: For simplicity, take the snapshot at $t = 0$ s so the equation becomes $y(x) = (3.0 \text{ cm})\cos(1.5x)$.

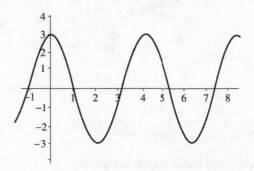

First compare Equation 15.7 for a wave moving to the right to our wave function to identify λ and T.

$$y(x,t) = A \cos\left(2\pi\left(\frac{x}{\lambda} - \frac{t}{T} \right) \right)$$

$$y(x,t) = (3.0 \text{ cm}) \cos(1.5x - 50t)$$

We see that $\lambda = (2\pi/1.5)$ m, $T = (2\pi/50)$ s, and (unneeded) $A = 3.0$ cm.

Now use Equation 15.9 for the wave speed.

$$v = \frac{\lambda}{T} = \frac{(2\pi/1.5) \text{ m}}{(2\pi/50) \text{ s}} = 33 \text{ m/s}$$

As mentioned, because of the minus sign in Equation 15.7 (and our wave function), the wave is moving to the right. **Assess:** We do not know what kind of wave this is, but the answer is reasonable for the wave speed on a string, and the amplitude is a reasonable amplitude for a wave on a string. Note that although we had to concern ourselves with the 2π, it did cancel out in our last calculation.

P15.69. Prepare: The presence of the wall has nothing to do with the intensity. The wall allows you to see the light, but the light wave has the same intensity at *all* points 2.0 m from the bulb whether it is striking a surface or moving through empty space. At a distance r from the bulb, the one watt of visible light has spread out to cover the surface of a sphere of radius r. The surface area of a sphere is $a = 4\pi r^2$. We will now use Equation 15.12 to find the intensity. On the other hand, the full two watts of light is concentrated in a dot of area $a = \pi r^2 = \pi(0.001 \text{ m})^2 = 3.14 \times 10^{-6}$ m^2.

Solve: (a) The intensity at a distance of 2 m is

$$I = \frac{P}{a} = \frac{P}{4\pi r^2} = \frac{4.0 \text{ W}}{4\pi(2.0 \text{ m})^2} = 0.080 \text{ W/m}^2$$

(b) Unlike the light from a light bulb, a laser beam does *not* spread out. We ignore the small diffraction spread of the beam. The laser beam creates a dot of light on the wall that is 2 mm in diameter. The full two watts of light is concentrated in this dot of area $a = \pi r^2 = \pi(0.001 \text{ m})^2 = 3.14 \times 10^{-6}$ m^2. The intensity is

$$I = \frac{P}{a} = \frac{2 \text{ W}}{3.14 \times 10^{-6} \text{ m}^2} = 640\,000 \text{ W/m}^2$$

Assess: Although the *power* of the laser is half the power of the light bulb, the laser produces light on the wall whose *intensity* is about eight million times that of the light bulb.

P15.75. Prepare: We are given the intensity level of the sound but we need the intensity so we solve Equation 15.14, $I = I_0 \times 10^{(\beta/10 \text{ dB})} = 1.0 \times 10^{-13}$ W/m^2. Let r_1 and I_1 be the distance to and intensity experienced by you and let r_2 and I_2 be the distance to and intensity experienced by the mouse. Then we can use Equation 15.13 to get the distance from the leaf to the mouse, r_2.

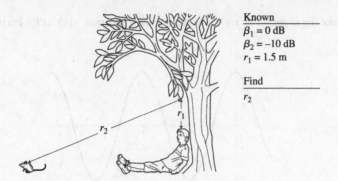

Known
$\beta_1 = 0$ dB
$\beta_2 = -10$ dB
$r_1 = 1.5$ m

Find
r_2

Solve: Solving Equation 15.13 for r_2, we obtain the following:

$$r_2 = \sqrt{r_1^2\left(\frac{I_1}{I_2}\right)} = \sqrt{(1.5 \text{ m})^2\left(\frac{1.0\times10^{-12} \text{ W/m}^2}{1.0\times10^{-13} \text{ W/m}^2}\right)} = 4.7 \text{ m}$$

Assess: To decrease β by 10 dB, the intensity must be decreased by a factor of 10. We know that intensity is inversely proportional to the distance to the source squared, so a reduction in the intensity by a factor of 10 means an increase in r by a factor of $\sqrt{10}$, which is about 3. So it makes sense that the mouse is about three times farther away from the leaf than you are.

P15.77. Prepare: The sound generator's frequency is altered by the Doppler effect. According to Equations 15.16, the frequency increases as the generator approaches the student, and it decreases as the generator recedes from the student. Convert rpm into SI units. Use 343 m/s for the speed of sound.
Solve: The generator's speed is

$$v_s = r\omega = r(2\pi f) = (1.0 \text{ m})2\pi\left(\frac{100}{60} \text{ rev/s}\right) = 10.47 \text{ m/s}$$

The frequency of the approaching generator is

$$f_+ = \frac{f_s}{1 - v_s/v} = \frac{600 \text{ Hz}}{1 - \frac{10.47 \text{ m/s}}{343 \text{ m/s}}} = 620 \text{ Hz}$$

Doppler effect for the receding generator, on the other hand, is

$$f_- = \frac{f_s}{1 + v_s/v} = \frac{600 \text{ Hz}}{1 + \frac{10.47 \text{ m/s}}{343 \text{ m/s}}} = 580 \text{ Hz}$$

Thus, the highest and the lowest frequencies heard by the student are 620 Hz and 580 Hz.

SUPERPOSITION AND STANDING WAVES

Q16.1. Reason: Where there is a change in medium—in particular a change in the wave speed—then reflection can occur.
Assess: Light travels at different speeds in water and air, and so some is reflected at a water-air interface.

Q16.5. Reason: The frequency of vibration of your vocal cords is related to their linear mass density by $f_m = (m/(2L))\sqrt{T/\mu}$. Due to the inflammation, the vocal cords are more massive and hence have a greater linear mass density. Since the inflamed vocal cords have a greater linear mass density, the frequency of vibration of the cords and hence the frequency of the sound generated will decrease.
Assess: You are no doubt aware that the frequency of your voice lowers with certain illnesses.

Q16.7. Reason: (a) When standing waves are set up in a tube that is open at both ends, the length of the tube is an integral number of half wavelengths $L = m\lambda/2$. Looking at the figure we see four half wavelengths, hence $m = 4$, and the air column is vibrating in the fourth harmonic. **(b)** Since this is sound, we have a longitudinal wave and the air molecules are vibrating horizontally, parallel to the tube.
Assess: The wave diagram superimposed on top of the open-ended tube sketch is a representation of the pressure at different points in the gas. It is not an air molecule displacement sketch.

Q16.11. Reason: Treating the instrument as an open-closed tube, the frequencies are $f_m = m(v/(4L))$. Inspecting this relationship, we expect the pitch of the instrument to be slightly higher since the speed of sound is greater in helium.
Assess: Having the ability to inspect a function such as the one previously discussed for the frequency in order to determine what will occur is a skill that physics will help you develop.

Q16.13. Reason: The introduction of helium into the mouth allows harmonics of higher frequencies to be excited more than in the normal voice. The fundamental frequency of the voice is the same but the quality has changed. Our perception of the quality is a function of which harmonics are present.
Assess: The fundamental frequency of a complex tone from the voice is determined by the vibration of the vocal cords and depends on the tension and linear mass density of the cords, not the gas in the mouth.

Q16.15. Reason: The settings on the synthesizer change the amount and proportion of higher harmonics that are mixed in (added to) the fundamental frequency.
Assess: Adding a lot of 2nd and 4th harmonics makes it sound more like a flute.

Q16.21. Reason: The maximum displacement will be the sum of the contributions from the two traveling waves at each point and at each time; however, you are not guaranteed to be watching a point where a crest will meet a crest. It is true that you might be watching an antinode where the maximum displacement would be $2A$, but it is also possible that you might be watching a node that doesn't move at all—or any place in between.
The correct choice is D.

Assess: If you watch the whole string, you will see points whose maximum displacement is $2A$ and other points whose maximum displacement is 0 and other points in between.

Q16.23. Reason: We know that the speed, frequency, and wavelength of a traveling wave are related by $f = v/\lambda$. As the air and pipe warm from $20°C$ to $25°C$, there is an insignificant expansion of the pipe, hence the resonant wavelengths remain the same. However, the change in temperature of the air is significant compared to the temperature of the air. Since the speed of sound increases with temperature, the frequency will increase. The correct response is A.
Assess: Your ability to look at an expression that describes a situation and predict what will happen under certain circumstances will be enhanced by your study of physics.

Q16.25. Reason: For the lowest standing wave mode on a string, the wavelength is twice the length of the string. If this is the case the speed of the disturbance is determined by
$$v = f_1 \lambda_1 = (20 \text{ Hz})(2.0 \text{ m}) = 40 \text{ m/s}$$
The correct choice is D.
Assess: For the lowest standing wave mode on a string, one-half a wavelength fits into the string.

Problems

P16.3. Prepare: The principle of superposition comes into play whenever the waves overlap.
Solve: As graphically illustrated in the following figure, the snapshot graph of Figure P16.3 was taken at $t = 4$ s.

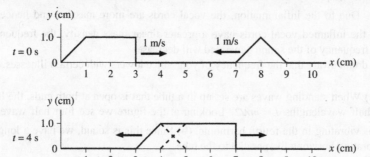

Assess: This is an excellent problem because it allows you to see the progress of each wave and the superposition (addition) of the waves. As time progresses, you know exactly what has happened to each wave and to the superposition of these waves.

P16.5. Prepare: The principle of superposition comes into play whenever the waves overlap. The waves are approaching each other at a speed of 1 m/s, that is, each part of each wave is moving 1 m every second.
Solve: The graph at $t = 1$ s differs from the graph at $t = 0$ s in that the left wave has moved to the right by 1 m and the right wave has moved to the left by 1 m. This is because the distance covered by the wave pulse in 1 s is 1 m. The snapshot graphs at $t = 2$ s, 3 s, 4 s, 5 s, and 6 s are a superposition of the left- and the right-moving waves. The overlapping parts of the two waves are shown by the dotted lines.

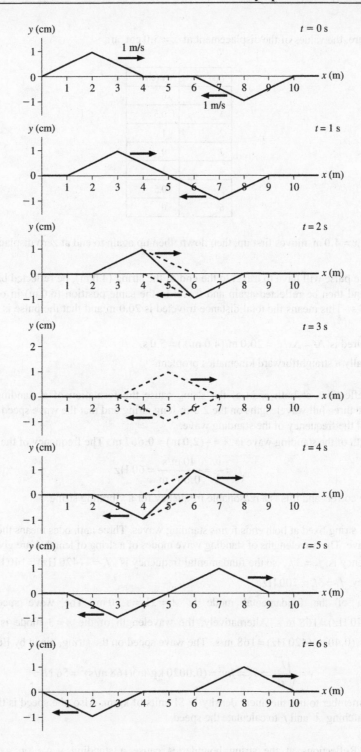

Solve: From the figure, the values of the displacement at $x = 5.0$ cm are

t (s)	y (cm)
0	0
1	0.5
2	1
3	0
4	–1
5	–0.5
6	0

Assess: The point at $x = 4.0$ m moves first up, then down, then up again to end at zero displacement.

P16.7. Prepare: The pulse will have to travel to the end of the string (4.0 m), be reflected back to the other end of the string (10.0 m), and then be reflected again and travel to the same position (6.0 m) in order to have the same appearance as at $t = 0$ s. This means the total distance traveled is 20.0 m and that the pulse is traveling at a speed of 4.0 m/s.
Solve: The time required is $\Delta t = \Delta x / v_x = 20.0$ m/(4.0 m/s) $= 5.0$ s.
Assess: This is basically a straightforward kinematics problem.

P16.9. Prepare: Reflections at both ends of the string cause the formation of a standing wave. Figure P16.9 indicates that there are three full wavelengths on the 2.0-m-long string and that the wave speed is 40 m/s. We will use Equation 15.10 to find the frequency of the standing wave.
Solve: The wavelength of the standing wave is $\lambda = \frac{1}{3}(2.0 \text{ m}) = 0.667$ m. The frequency of the standing wave is thus

$$f = \frac{v}{\lambda} = \frac{40 \text{ m/s}}{0.667 \text{ m}} = 60 \text{ Hz}$$

Assess: The units are correct and this is a reasonable frequency for a vibrating string.

P16.15. Prepare: A string fixed at both ends forms standing waves. Three antinodes means the string are vibrating as the $m = 3$ standing wave. The wavelengths of standing wave modes of a string of length L are given by Equation 16.1.
Solve: (a) The frequency is $f_3 = 3f_1$, so the fundamental frequency is $f_1 = \frac{1}{3}(420 \text{ Hz}) = 140$ Hz. The fifth harmonic will have the frequency $f_5 = 5f_1 = 700$ Hz.
(b) The wavelength of the fundamental mode is $\lambda_1 = 2L = 1.20$ m. The wave speed on the string is $v = \lambda_1 f_1 = (1.20 \text{ m})(140 \text{ Hz}) = 168$ m/s. Alternatively, the wavelength of the $n = 3$ mode is $\lambda_3 = \frac{1}{3}(2L) = 0.40$ m, from which $v = \lambda_3 f_3 = (0.40 \text{ m})(420 \text{ Hz}) = 168$ m/s. The wave speed on the string, given by Equation 15.2, is

$$v = \sqrt{\frac{T_S}{\mu}} \Rightarrow T_S = \mu v^2 = (0.0020 \text{ kg/m})(168 \text{ m/s})^2 = 56 \text{ N}$$

Assess: You must remember to use the linear density in SI units of kg/m. Also, the speed is the same for all modes, but you must use a matching λ and f to calculate the speed.

P16.17. Prepare: Reflections at the string boundaries cause a standing wave on a stretched string. The wavelengths of standing wave modes of a string of length L are given by Equation 16.1, so we can easily determine the wavelength from the vibrating length of the string, which is 1.90 m. With a known frequency of 27.5 Hz we can find the wave speed using Equation 15.10. Equation 15.2 will now allow us to find the tension in the wire. Mass density μ of the wire is equal to the ratio of its mass and length.
Solve: Because the vibrating section of the string is 1.9 m long, the two ends of this vibrating wire are fixed, and the string is vibrating in the fundamental harmonic. The wavelength is

$$\lambda_m = \frac{2L}{m} \Rightarrow \lambda_1 = 2L = 2(1.90\,\text{m}) = 3.80\,\text{m}$$

The wave speed along the string is $v = f_1\lambda_1 = (27.5\,\text{Hz})(3.80\,\text{m}) = 104.5\,\text{m/s}$. The tension in the wire can be found as follows:

$$v = \sqrt{\frac{T_S}{\mu}} \Rightarrow T_S = \mu v^2 = \left(\frac{\text{mass}}{\text{length}}\right) v^2 = \left(\frac{0.400\,\text{kg}}{2.00\,\text{m}}\right)(104.5\,\text{m/s})^2 = 2180\,\text{N}$$

Assess: You must remember to use the linear density in SI units of kg/m. Also, the speed is the same for all modes, but you must use a matching λ and f to calculate the speed.

P16.21. Prepare: We want to use the fundamental relationship for periodic waves. But first, convert the length to SI units (use Table 1.3).

$$L = 18\,\text{ft}\left(\frac{0.30\,\text{m}}{1\,\text{ft}}\right) = 5.49\,\text{m}$$

Solve: (a) For an open-closed pipe in the fundamental mode $L = \frac{1}{4}\lambda$ (one-quarter of a wavelength fits in the pipe),

$$\lambda = 4L = 4(5.49\,\text{m}) = 21.9\,\text{m}$$

$$f = \frac{v}{\lambda} = \frac{350\,\text{m/s}}{21.9\,\text{m}} = 16\,\text{Hz}$$

This is below the arbitrary lower limit of the range of human hearing.
(b) We notice that the true value (27.5 Hz) is different from the answer we got in part **(a)**. To find the "effective length" of the instrument with a fundamental frequency of 27.5 Hz using the open-closed tube model, we simply do the previous calculations in reverse order. First find λ for the fundamental mode.

$$\lambda = \frac{v}{f} = \frac{350\,\text{m/s}}{27.5\,\text{Hz}} = 12.7\,\text{m}$$

Now

$$L = \frac{1}{4}\lambda = \frac{1}{4}(12.7\,\text{m}) = 3.18\,\text{m} = 10.4\,\text{ft} \approx 10\,\text{ft}$$

Assess: The "effective length" is just over half of the real length. This would lead us to conclude that the open-closed tube model is not a very accurate model for this situation. A good contrabassoon gives foundation and body to the orchestra.

P16.23. Prepare: For the open-open tube, the two open ends exhibit antinodes of a standing wave. The possible wavelengths for this case are $\lambda_m = 2L/m$, where $m = 1, 2, 3 \ldots$ On the other hand, in the case of an open-closed tube $\lambda_m = 4L/m$, where $m = 1, 3, 5 \ldots$ The length of the tube is 121 cm.
Solve: (a) The three longest wavelengths are

$$\lambda_1 = \frac{2(1.21\,\text{m})}{1} = 2.42\,\text{m} \qquad \lambda_2 = \frac{2(1.21\,\text{m})}{2} = 1.21\,\text{m} \qquad \lambda_3 = \frac{2(1.21\,\text{m})}{3} \approx 0.807\,\text{m}$$

(b) The three longest wavelengths are

$$\lambda_1 = \frac{4(1.21\,\text{m})}{1} = 4.84\,\text{m} \qquad \lambda_2 = \frac{4(1.21\,\text{m})}{3} = 1.61\,\text{m} \qquad \lambda_3 = \frac{4(1.21\,\text{m})}{5} \approx 0.968\,\text{m}$$

Assess: It is clear that the end of the air column, whether open or closed, changes the possible modes.

P16.27. Prepare: The frequencies at which resonance will occur in an open-open pipe are determined by $f_m = m(v/(2L))$.
Solve: (a) Knowing the speed of sound, the fundamental frequency is determined by
$$f_1 = (1)(v/(2L)) = (340\,\text{m/s})/(2(30.0\,\text{m})) = 5.67\,\text{Hz}$$

(b) The lowest frequency we can hear is about 20 Hz. For an open-open tube the frequency of the harmonics are related to the fundamental frequency by $f_m = mf_1$, hence the lowest harmonic that would be audible to the human ear is

$$m = f_m/f_1 = 20 \text{ Hz}/(5.67 \text{ Hz}) = 3.5$$

But m must be an integer, so we must take $m = 4$ and calculate the frequency.

$$4f_1 = 4(5.67 \text{ Hz}) = 22.7 \text{ Hz}$$

(c) As the air cools the speed of sound will decrease. Inspecting the function given in the previous Prepare step, we see that this would in turn result in a decrease in frequency.

Assess: This is a straightforward example of resonance in an open-open pipe.

P16.29. Prepare: For an open-closed tube we need Equation 16.7

$$f_m = m\frac{v}{4L}$$

where we are given that $m = 1$ (for the fundamental frequency) and that $f_1 = 200$ Hz. We are also given that $v = 350$ m/s.

Solve: Solve the equation for L.

$$L = m\frac{v}{4f_m} = (1)\frac{350 \text{ m/s}}{4(200 \text{ Hz})} = 0.44 \text{ m} = 44 \text{ cm}$$

Assess: Since the 200 Hz is a "typical" fundamental frequency we don't really expect the length obtained to be the exact length of *your* vocal tract; but we do note that it is in the right ballpark (put a meter stick next to your stretched neck and guesstimate the distance from your mouth to your diaphragm).

P16.31. Prepare: Follow Example 16.6 very closely. Assume the ear canal is an open-closed tube, for which we need Equation 16.7

$$f_m = m\frac{v}{4L} \qquad m = 1, 3, 5, 7 \ldots$$

where we are given that $L = 1.3$ cm $= 0.013$ m. We assume that $v = 350$ m/s in the warm ear canal.

Take the audible range as 20 Hz–20 kHz.

Solve: Plug in various values of m and obtain the corresponding frequencies.

$$f_1 = (1)\frac{350 \text{ m/s}}{4(0.013 \text{ m})} = 6730 \text{ Hz} \approx 6700 \text{ Hz}$$

Higher frequencies are odd multiples of this fundamental.

$$f_3 = 3(6730 \text{ Hz}) = 20200 \text{ Hz}$$

Already f_3 is out of the audible range. So the only one in the audible range is $f_1 = 6700$ Hz.

P16.35. Prepare: First assume the speed of sound is 350 m/s in the vocal tract. The relationship between speed, wavelength, and frequency for a traveling wave disturbance in any medium is $v = f\lambda$. The frequency of vibration in air is caused by and is the same as the frequency of vibration of the vocal cords. The length of the vocal tract is an integral number of half-wavelengths $L = m\lambda/2$. The length of the vocal tract and hence the wavelengths that cause standing wave resonance do not change as the diver descends. However, since the speed of the sound waves changes, the frequency will also change.

Solve: When a sound of frequency 270 Hz is coming out of the vocal tract, the wavelength of standing waves established in the vocal tract associated with this frequency is

$$\lambda = v/f = (350 \text{ m/s})/(270 \text{ Hz}) = 1.296 \text{ m}$$

As the diver descends, the vocal tract does not change length and hence this wavelength for standing wave resonance will not change. However since the sound is now travelling through a helium-oxygen mixture with a speed of 750 m/s, the frequency of the sound will change to

$$f = v/\lambda = (750 \text{ m/s})/(1.296 \text{ m}) = 580 \text{ Hz}$$

Going through the same procedure for sound at a frequence of 2300 Hz we get the frequency in the helium-oxygen mixture to be $f = 4900\,\text{Hz}$.

Assess: We are aware that the sound should be at a higher frequency and the frequencies obtained have higher values.

P16.39. Prepare: We assume that the speakers are identical and that they are emitting in phase. Since you don't hear anything, the separation between the two speakers corresponds to the condition of destructive interference.
Solve: Equation 16.9 for destructive interference is

$$\Delta d = \left(m + \frac{1}{2}\right)\lambda \Rightarrow \Delta d = \frac{\lambda}{2}, \frac{3\lambda}{2}, \frac{5\lambda}{2}$$

Since the wavelength is

$$\lambda = \frac{v}{f} = \frac{340\,\text{m/s}}{170\,\text{Hz}} = 2.0\,\text{m}$$

three possible values for d are 1.0 m, 3.0 m, and 5.0 m.
Assess: The units worked out and these are reasonable distances.

P16.41. Prepare: The two speakers are identical, and so they are emitting circular waves in phase. The overlap of these waves causes interference. An overview of the problem follows.

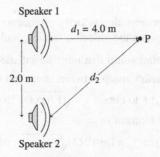

Speaker 1

$d_1 = 4.0\,\text{m}$

P

2.0 m

d_2

Speaker 2

Solve: The wavelength of the sound waves is

$$\lambda = \frac{v}{f} = \frac{340\,\text{m/s}}{1800\,\text{Hz}} = 0.1889\,\text{m}$$

From the geometry of the figure,

$$d_2 = \sqrt{d_1^2 + (2.0\,\text{m})^2} = \sqrt{(4.0\,\text{m})^2 + (2.0\,\text{m})^2} = 4.472\,\text{m}$$

So, $\Delta d = d_2 - d_1 = 4.472\,\text{m} - 4.0\,\text{m} = 0.472\,\text{m}$.
Because $\Delta d/\lambda = 0.472\,\text{m}/0.1889\,\text{m} = 2.5 = 5/2 = 2 + \frac{1}{2}$ or $\Delta d = 2 + \frac{1}{2}\lambda$, the interference is perfectly destructive.
Assess: Destructive interference (for two waves in phase) will occur when the path difference is an integral number of half wavelengths.

P16.43. Prepare: The beat frequency is the difference of the two frequencies. We know the flat flute's frequency is lower.
Solve: $f_{\text{untuned}} = f_{\text{tuned}} - f_{\text{beat}} = 440\,\text{Hz} - 2\,\text{Hz} = 438\,\text{Hz}$.
Assess: Trained ears are quite sensitive to such differences in frequency.

P16.47. Prepare: Knowing the expression for the beat frequency, we can determine by which the amount the frequency will change. But at this point we don't know if the frequency increases or decreases. Examining the expression for the frequency of a flute (modeled as an open-open pipe) as a function of its length, we can establish if the frequency increases or decreases when the "tuning joint" is removed.
Solve: Using the expression for the beat frequency, the flute player's initial frequency is either $523\,\text{Hz} + 4\,\text{Hz} = 527\,\text{Hz}$ or $523\,\text{Hz} - 4\,\text{Hz} = 519\,\text{Hz}$. Modeling the flute as an open-open pipe we see that $v = f\lambda$ and $L = m\lambda/2$, which may be combined to obtain $f = mv/(2L)$

This expression allows us to see that as the length increases, the frequency decreases. As a result we know that the initial frequency of sound from the flute was 527 Hz.

Assess: Since she matches the tuning fork's frequency by lengthening her flute, she is increasing the wavelength of the standing wave in the flute. A wavelength increase means a decrease of frequency because $v = f\lambda$. This confirms that the initial frequency was greater than the frequency of the tuning fork.

P16.49. Prepare: Since when a sound wave hits the boundary between soft tissue and air, or between soft tissue and bone, most of the energy is reflected, the situation is like a vibrating string with reflections at both ends. If that is the case, standing wave resonance will be established (and heating will occur) when twice the thickness of the soft tissue is an integral number of wavelengths. The speed of ultrasound in soft tissue is 1540 m/s.

Solve: For a standing wave,

$$f_m = m\left(\frac{v}{2L}\right) \Rightarrow m = \frac{f_m(2L)}{v} = \frac{(0.70\times10^6 \text{ Hz})(2)(0.0055 \text{ m})}{1540 \text{ m/s}} = 5.0$$

Since this is an integer, then yes, there will be standing waves and heating of the tissue.

Assess: Since the frequency of the ultrasound and the thickness of the soft tissue are reasonable, we expect heating to occur.

P16.51. Prepare: The relationship between the velocity, frequency, and wavelength of a traveling wave disturbance is $v = f\lambda$. The relationship between the velocity, tension, and linear mass density for a traveling wave disturbance in a string is $v = \sqrt{T/\mu}$. The relationship that must be satisfied to create standing wave resonance in a stretched tendon is $L = m\lambda/2$. Finally, the relationship between linear and volume mass density is $\mu = \rho A$.

Solve: Start with $v = \sqrt{T/\mu}$ and insert $\mu = \rho A$ to obtain $v = \sqrt{T/(\rho A)}$. Express the wavelength as $\lambda = 2L/m$. Insert both v and λ into $v = f\lambda$ and solve for the frequency.

$$\sqrt{T/(\rho A)} = f_m(2L/m) \Rightarrow f_m = (m/(2L))\sqrt{T/(\rho A)} \quad \text{where } m = 1, 2, 3, \ldots$$

Inserting values, obtain the fundamental frequency:

$$f_1 = (1/(2(0.20 \text{ m})))\sqrt{500\,\text{N}/((1100 \text{ kg/m}^3)(1.00\times10^{-4} \text{ m}^2))} = 160 \text{ Hz}$$

Other possible frequencies are multiples of the fundamental
$$f_2 = 2f_1 = 320 \text{ Hz}, \quad f_3 = 3f_1 = 480 \text{ Hz}, \ldots \text{ etc.}$$

Assess: These are reasonable frequencies for the vibration of a tendon.

P16.55. Prepare: For the stretched wire vibrating at its fundamental frequency, the wavelength of the standing wave from Equation 16.1 is $\lambda_1 = 2L$. From Equation 15.2, the wave speed is equal to $\sqrt{T_S/\mu}$, where $\mu = \text{mass/length} = 5.0\times10^{-3} \text{ kg}/0.90 \text{ m} = 5.555\times10^{-3} \text{ kg/m}$. The tension T_S in the wire equals the weight of the sculpture or Mg.

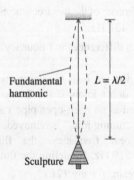

Fundamental harmonic $L = \lambda/2$

Sculpture

Solve: The wave speed on the steel wire is

$$v_{wire} = \sqrt{\frac{T_s}{\mu}} = \sqrt{\frac{Mg}{\mu}} = \sqrt{\frac{(12 \text{ kg})(9.8 \text{ m/s}^2)}{5.55 \times 10^{-3} \text{ kg/m}}} = 145.6 \text{ m/s}$$

Now we can solve for frequency.

$$v_{wire} = f\lambda \Rightarrow f = \frac{v_{wire}}{\lambda} = \frac{v_{wire}}{2L} = \frac{145.6 \text{ m/s}}{2(0.90 \text{ m})} = 81 \text{ Hz}$$

Assess: A frequency of 81 Hz for the wire is reasonable.

P16.59. Prepare: The nodes of a standing wave are spaced $\lambda/2$ apart. The wavelength of the mth mode of an open-open tube from Equation 16.6 is $\lambda_m = 2L/m$. Or, equivalently, the length of the tube that generates the mth mode is $L = m(\lambda/2)$. Here λ is the same for all modes because the frequency of the tuning fork is unchanged.

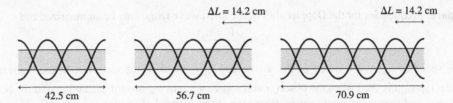

$\Delta L = 14.2$ cm $\Delta L = 14.2$ cm

42.5 cm   56.7 cm   70.9 cm

Solve: Increasing the length of the tube to go from mode m to mode $m + 1$ requires a length change:

$$\Delta L = (m+1)(\lambda/2) - m\lambda/2 = \lambda/2$$

That is, lengthening the tube by $\lambda/2$ adds an additional antinode and creates the next standing wave. This is consistent with the idea that the nodes of a standing wave are spaced $\lambda/2$ apart. This tube is first increased by $\Delta L = 56.7$ cm $- 4.25$ cm $= 14.2$ cm, then by $\Delta L = 70.9$ cm $- 56.7$ cm $= 14.2$ cm. Thus $\lambda/2 = 14.2$ cm and $\lambda = 28.4$ cm $= 0.284$ m. Therefore, the frequency of the tuning fork, using Equation 15.10, is

$$f = \frac{v}{\lambda} = \frac{343 \text{ m/s}}{0.284 \text{ m}} = 1210 \text{ Hz}$$

Assess: This is a reasonable value for the frequency of a tuning for k in the audible range and the units are correct.

P16.61. Prepare: The waves constructively interfere when speaker 2 is located at 0.75 m and 1.00 m, but not in between. Assume the two speakers are in phase (helpful for visualization, but the result will be generally true as long as the two frequencies are the same). For constructive interference the path length difference must be an integer number of wavelengths, 0.75 m $= n\lambda$, and 1.00 m $= (n+1)\lambda$. Subtracting the two equations gives $\lambda = 0.25$ m.

Solve:

$$f = \frac{v}{\lambda} = \frac{340 \text{ m/s}}{0.25 \text{ m}} = 1360 \text{ Hz} \approx 1400 \text{ Hz}$$

Assess: 1400 Hz is near the "middle" of the range of human hearing, so it is probably right.

P16.63. Prepare: The changing sound intensity is due to the interference of two overlapped sound waves. Minimum intensity implies destructive interference. Destructive interference occurs where the path length difference for the two waves is $\Delta d = (m + \frac{1}{2})\lambda$.

Solve: The wavelength of the sound is $\lambda = v_{sound}/f = (343 \text{ m/s})/(686 \text{ Hz}) = 0.500$ m. Consider a point that is a distance d in front of the top speaker. Let d_1 be the distance from the top speaker to the point and d_2 the distance from the bottom speaker to the point. We have

$$d_1 = x \quad d_2 = \sqrt{x^2 + (3.00 \text{ m})^2}$$

Destructive interference occurs at distances d such that

$$\Delta d = \sqrt{x^2 + 9 \text{ m}^2} - x = \left(m + \frac{1}{2}\right)\lambda$$

To solve for x, isolate the square root on one side of the equation and then square:

$$x^2 + 9 \text{ m}^2 = \left[x + \left(m + \frac{1}{2} \right) \lambda \right]^2 = x^2 + 2 \left(m + \frac{1}{2} \right) \lambda x + \left(m + \frac{1}{2} \right)^2 \lambda^2 \Rightarrow x = \frac{9 \text{ m}^2 - (m + \frac{1}{2})^2 \lambda^2}{2(m + \frac{1}{2})\lambda}$$

Evaluating x for different values of m:

m	x (m)
0	17.88
1	5.62
2	2.98
3	1.79

Because you start at $x = 2.5$ m and walk *away* from the speakers, you will only hear minima for values $x > 2.5$ m. Thus, minima will occur at distances of 2.98 m, 5.62 m, and 17.88 m.

Assess: These are reasonable distances and the units are correct.

P16.67. Prepare: Frequencies for the Doppler shift in the microwave range may be summarized by:

$$f_\pm = f_s \frac{(c \pm v_0)}{(c \mp v_s)}$$

Where $c = $ the speed of light, $v_s = $ speed of the source and $v_0 = $ speed of the observer. In the numerator and the denominator, the top sign is used when the observer and source are moving toward each other and the bottom sign is used if the observer and source are moving away from each other. A speed of 55 mph is essentially 25 m/s.

Solve: The frequency sent out by the radar unit is $f_s = 10.5 \times 10^9$ Hz.

The frequency of waves observed by the moving car is

$$f_c = f_s \frac{(c + v_c)}{c} = 10.5 \times 10^9 \text{ Hz} \frac{(3.00 \times 10^8 + 25) \text{ m/s}}{3.00 \times 10^8 \text{ m/s}} = 10.50000088 \times 10^9 \text{ Hz}$$

Now the car acts like a source of this frequency and since it is moving towards the police unit, the frequency of the reflected waves arriving at the unit may be determined by

$$f_u = f_c \frac{c}{c - v_c} = (10.50000088 \times 10^9 \text{ Hz}) \frac{3.00 \times 10^8 \text{ m/s}}{(3.00 \times 10^8 - 25) \text{ m/s}} = 10.500001750 \times 10^9 \text{ Hz}$$

As a result, the beat frequency determined by the unit is 1750 Hz.

When the unit is switched to calibration mode, a tuning fork vibrating in front of the unit at a frequency of 1750 Hz will register on the unit as 55 mph.

Assess: A frequency of 1750 Hz for a tuning fork is in the range that we hear very well. According to the operations manual that comes with the police radar unit, to ensure accuracy every unit is supposed to be calibrated with tuning forks before each working shift. Using several different tuning forks would ensure that the unit is accurate over a range of speeds. Finally, the problem could be solved in one step by inserting values into the first equation, but it is instructive to solve the problem in two steps as shown previously.

P16.69. Prepare: We will need concepts from Chapter 15 to solve this problem. The speed of ultrasound waves in human tissue is given in Table 15.1 as $v = 1540$ m/s. The frequency of the reflected wave must be 2.0 MHz $\pm$ 520 Hz, that is, $f_+ = 2000520$ Hz and $f_- = 1999480$ Hz.

There are a couple of mathematical paths we could take, but it is probably easiest to carefully review Example 15.13 and use the mathematical result there, as it is very similar to our problem and gives an expression for the speed of the source v_s.

Solve:

$$v_s = \frac{f_+ - f_-}{f_+ + f_-} v = \frac{2000520 \text{ Hz} - 1999480 \text{ Hz}}{2000520 \text{ Hz} + 1999480 \text{ Hz}} 1540 \text{ m/s} = \frac{1040 \text{ Hz}}{4000000 \text{ Hz}} 1540 \text{ m/s} = 0.40 \text{ m/s}$$

Assess: We may not have a good intuition about how fast heart muscles move, but 0.40 m/s seems neither too fast nor too slow for a maximum speed.

PART IV

OSCILLATIONS AND WAVES

PptIV.7. Reason: Use $T = 1/f$ in $v = \lambda f$ and the given equation for v in deep water. The effective speed with which the waves pass the ship is the speed of the waves plus the speed of the ship toward the waves.

$$T = \frac{\lambda}{v} = \frac{\lambda}{\sqrt{\dfrac{g\lambda}{2\pi}} + v_{\text{ship}}} = \frac{75\,\text{m}}{\sqrt{\dfrac{(9.8\,\text{m/s}^2)(75\,\text{m})}{2\pi}} + 4.5\,\text{m/s}} = 4.9\,\text{s} \approx 5\,\text{s}$$

The answer is B.

Assess: We expect the waves to reach the ship more frequently if the ship is sailing toward them.

PptIV.9. Reason: We need to know how far 40 wavelengths is, so we find λ. In Table 15.1 find the speed of ultrasound in human tissue: $v = 1540\,\text{m/s}$.

$$\lambda = \frac{v}{f} = \frac{1540\,\text{m/s}}{1.0\,\text{MHz}} = 0.00154\,\text{m}$$

Now find the number of wavelengths in 12 cm.

$$12\,\text{cm} = 12\,\text{cm}\left(\tfrac{1\,\text{wavelength}}{0.00154\,\text{m}}\right) = 78 \text{ wavelengths.}$$

This is about twice the 40-wavelength half-distance, so the intensity is halved twice. In other words, the new intensity is 250 W/m². The answer is C.

Assess: The penetration depth depends on the frequency, but this is reasonable for 1 MHz ultrasound.

PptIV.11. Reason: Higher frequency provides less penetration, as we saw in the previous problem, but with smaller wavelength it provides better resolution. See Example 15.7. The correct answer is B.

Assess: There are tradeoffs in deciding which frequency to use.

PptIV.13. Reason: There are three full oscillations in 0.004 s so the period is 0.00133 s. The frequency is the inverse of this.

$$f = \frac{1}{T} = \frac{3}{0.004\,\text{s}} = 750\,\text{Hz}$$

The answer is C.

Assess: All of the choices are plausible answers, so we must have faith in the math.

PptIV.17. Reason: It will take half a period to return to the starting spot.

$$\frac{T}{2} = \pi\sqrt{\frac{L}{g}} = \pi\sqrt{\frac{5.0\,\text{m}}{9.8\,\text{m/s}^2}} = 2.2\,\text{s}$$

The answer is C.

Assess: This seems about right for a 5.0 m long swing.

PptIV.19. Reason: At the farthest point you are 1.0 m higher than at the bottom. See figure.

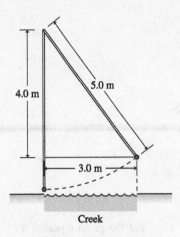

Creek

We now use conservation of energy to find the speed at the bottom.

$$\frac{1}{2}mv^2 = mgh \implies v = \sqrt{2gh} = \sqrt{2(9.8 \text{ m/s}^2)(1.0 \text{ m})} = 4.4 \text{ m/s}$$

The correct answer is C.

Assess: This seems like a reasonable speed for a big swing.